Lecture Notes in Geoinformation and Cartography

Series Editors: William Cartwright, Georg Gartner, Liqiu Meng, Michael P. Peterson

Poh C. Lai • Ann S.H. Mak
(Eds.)

GIS for Health and the Environment

Development in the Asia-Pacific Region

With 110 Figures

Springer

Editors:

Poh C. Lai
Department of Geography
The University of Hong Kong
Hong Kong Special Administrative
Region, China

Ann S.H. Mak
ERM Hong Kong
Taikoo Place, Island East
Hong Kong Special Administrative
Region, China

ISSN 1863-2246
ISBN 978-3-642-09040-0
ISBN 978-3-540-71318-0 (eBook)

Springer is a part of Springer Science+Business Media
springeronline.com

Cover design: deblik, Berlin

Preface

> "As the world becomes more integrated through the trade of goods and services and capital flows, it has become easier for diseases to spread through states, over borders and across oceans — and to do serious damage to vulnerable human and animal populations."
> *American RadioWorks and NPR News, 2001*

The global cost of communicable diseases is expected to rise. SARS has put the world on alert. We have now Avian Flu on the watch. Recognizing the global nature of threats posed by new and re-emerging infectious diseases and the fact that many recent occurrences originated in the Asia Pacific regions, there has been an increased interest in learning and knowing about disease surveillance and monitoring progresses made in these regions. Such knowledge and awareness is necessary to reduce conflict, discomfort, tension and uneasiness in future negotiations and global cooperation.

Many people are talking about the GIS and public and environmental health. The way we make public policies on health and environmental matters is changing, and there is little doubt that GIS provides powerful tools for visualizing and linking data in public health surveillance. This book is a result of the International Conference in GIS and Health held on 27-29 June 2006 in Hong Kong. The selected chapters are organized into four themes: GIS Informatics; Human and Environmental Factors; Disease modeling; and Public health, population health technologies, and surveillance.

As evident from the chapters, the main problem in GIS-based epidemiological studies is the availability of reliable exposure data. There is also a huge problem of showing adequate responsibility and ability to meet public concerns, such as protection on privacy and quick response systems. There has been some works done in search of the right approach in bringing together and reconciling market and public interests. Talking to each other and sharing critical information are getting increasingly important. Much work remains to be done to improve the GIS-based epidemiologic methods into tools for fully developed analytical studies and, particularly, the need to identify standard interfaces and infrastructures for the global disease reporting system.

January 2007

Poh C. Lai
Ann S.H. Mak

International Conference in GIS and Health 2006

Geospatial Research and Application Frontiers in Environmental and Public Health Systems [1]

Conference Chair

Poh C. Lai, University of Hong Kong, China

Program Committee

<u>*International Members*</u>

Chuleeporn Jiraphongsa, Ministry of Public Health, Thailand
Nina Lam, Louisiana State University, USA
Feng Lu, Chinese Academy of Sciences, China
Augusto Pinto, World Health Organization, France
Jan Rigby, University of Sheffield, United Kingdom
Pratap Singhasivanon, University of Mahidol, Thailand
Chris Skelly, Brunel University, United Kingdom

<u>*Local Members*</u>

Ping Kwong Au Yeung, Lands Department
Lorraine Chu, Mappa Systems Limited
Tung Fung, Chinese University of Hong Kong
Tai Hing Lam, University of Hong Kong
Hui Lin, Chinese University of Hong Kong
Christopher Hoar, NGIS China Limited
S.V. Lo, Health Welfare and Food Bureau
Ann Mak, ERM Company Limited
Stanley Ng, MapAsia Company Limited
Wenzhong Shi, Hong Kong Polytechnic University
Winnie Tang, ESRI China (Hong Kong) Limited
Raymond Wong, Intergraph Hong Kong
Anthony Gar-On Yeh, University of Hong Kong
Qiming Zhou, Hong Kong Baptist University

Executive Committee

Kawin K.W. Chan, University of Hong Kong
Richard K.H. Kwong, University of Hong Kong
Poh C. Lai, University of Hong Kong
Sharon T.S. Leung, NGIS China Limited
Feng Lu, Chinese Academy of Sciences
Ann S.H. Mak, ERM Company Limited
Franklin F.M. So, Experian Limited
Andrew S.F. Tong, University of Hong Kong

---

[1] The conference was a joint event held in June 2006 and jointly organized by the Department of Geography at the University of Hong Kong and the State Key Laboratory of Resources and Environmental Information Systems of the Chinese Academy of Sciences. It was supported by the Croucher Foundation and the Professional Services Development Assistance Scheme of the Commerce, Industry and Technology Bureau of the Government of Hong Kong.

Table of Contents

GIS Informatics

Exploratory Spatial Analysis Methods in Cancer Prevention and Control

Gerard Rushton

The University of Iowa

Abstract: Improved geocoding practices and population coverage of cancer incidence records, together with linkages to other administrative record systems, permit the development of new methods of exploratory spatial analysis. We illustrate these developments with results from a GIS-based workbench developed by faculty and students at the University of Iowa. The system accesses records from the Iowa Cancer Registry. In using these methods, the privacy of individuals is protected while still permitting results to be available for small geographic areas. Geographic masking techniques are illustrated as are kernel density estimation methods used in the context of Monte Carlo simulations of spatial patterns of selected cancer burdens of breast, colorectal and prostate cancer in Iowa.

Keywords: cancer prevention and control, exploratory spatial analysis

1 The need for maps in cancer prevention and control

The theme of this chapter is the design of cancer maps for cancer control and prevention activities. Abed et al. (2000) describe a framework for developing knowledge for making decisions for comprehensive cancer control and prevention. The decisions these authors have in mind involve local communities setting objectives, planning strategies, implementing them, and finally, determining improvements in health achieved by their activities. Each of these steps is explicitly spatial: where activities are directed, who is affected, and whose health is improved? Location is a critical part of this framework.

As with all chronic diseases, factors that influence the burden of the disease on any population include the behaviors of people, characteristics of environments, and availability and accessibility of health screenings and treatments. Objectives to improve population health, therefore, must identify spatial differences in these factors and must address strategies to change them in ways that will lead to improved health outcomes. Cancer maps play an important role in this process. Particularly geographic aspects of these tasks are:

- Spatial allocation of resources;
- Identification of areas with higher than expected incidence rates (disease clusters);
- Optimal location of services.

All three tasks require that the maps of the cancer burdens should capture any special demographic characteristics of local populations so that actions for control and prevention relate to population characteristics. None of these tasks should use cancer rates adjusted to standard population characteristics. Yet, these are precisely the characteristics of many cancer maps—see, for example, Pickle et al. 1996; Devesa et al. 1999.

1.1 The limitations of cancer mortality maps

In the short history of mapping cancer, most attention has been given to mapping cancer mortality; for most countries, cancer mortality data are collected routinely.

Since the geocode on a typical death certificate is some politically recognized area—often, in the United States a county—data is available for counties and most maps use counties or aggregates of counties, (Devesa et al. 1999). Mortality maps, however, are not so useful for planning control and prevention interventions because spatial variations in mortality rates can be due to differences in behaviors, in the environment or in local health system characteristics. Yet, untangling risks due to differences in these three factors is precisely what is required before plans to reduce cancer burdens can be established. With the development of cancer registries, however, data is available that allows attempts to be made to separate these influences and to develop interventions that will optimally reduce rates. Cancer maps have a vital role to play by mapping these factors, in addition to mortality.

1.2 The potential contribution of cancer registry data

There are two ways in which cancer registry data can be used for making cancer maps. They can be used to break down the burden of cancer on local populations into component parts. Assumed here is that the cancer registry is population-based; i.e. it accounts for all cases of cancer in a defined population. Although it may rely on health care facilities for much of its data, it must not be facility based. In most cases, registries are area-based and track down incidences of cancer in its defined population wherever they are diagnosed and treated. The components of interest are first confirmed diagnoses of cancer; the stage of the disease at the time of first diagnosis; the first course of treatment, survival rate, and mortality rate. Other components of cancer are screening rates and treatment rates. Data availability for these components often depends on the comprehensiveness of the health information available for the defined population (see Armstrong 1992).

1.3 The role of exploratory spatial analysis

In "exploratory spatial analysis" of cancer, geographic scale and pattern are explored. Each cancer map represents a decision to focus on a defined geographic scale and specific patterns may be revealed—or concealed—by the scale chosen. Figure 1 illustrates this principle using three infant mortality maps of one county in central Iowa. Approximately 20,000 births and 190 infant deaths occurred in this county in the four year period from 1989 through 1992. After geocoding each birth and death to its residential address, the three maps on the right of Figure 1 show the pattern at the scales captured by three, commonly used, administrative areas. A property of these maps is that the variability of the infant mortality rates depends on the size of the areas mapped. The rate for Zipcodes varies from 0 to 20 deaths per thousand births; for census tracts the rate varies from 0 to 36 and for census block groups the rate varies from 0 to 72. The legends for each map—not shown here—must necessarily be adjusted to accommodate these different variances. The sensitivity of the patterns of infant mortality to scale are clear on the left where geographic scales of the three maps are formally defined as spatial filters of 1.2, 0.8, and 0.4 miles respectively—applied in each case to a 0.4 mile grid from which the density estimates were made (see Bithell 1990; Rushton and Lolonis 1996). Again, on the left, patterns are different and depend on scale. We can conclude that patterns depend on scale and actions based on patterns should consider the scales at which the patterns were derived and ask whether the actions contemplated are reliably based on the data that supported them.

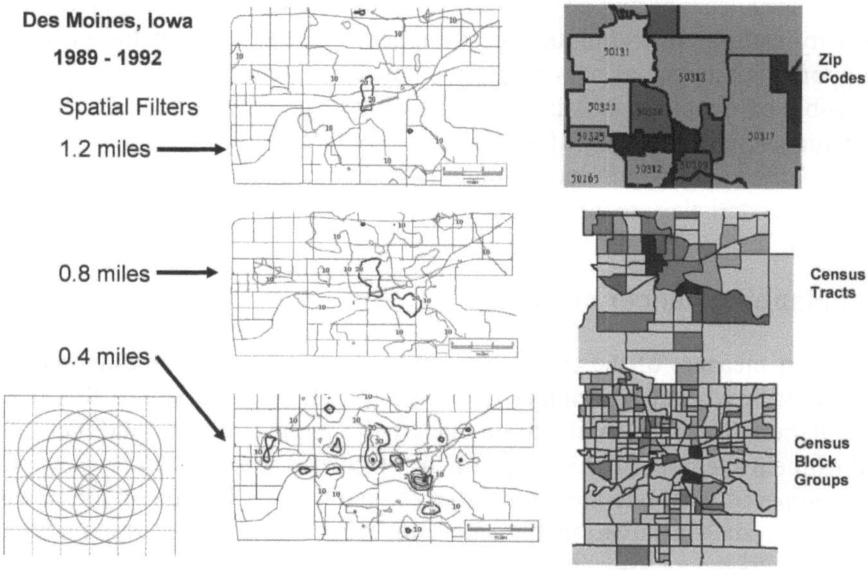

Fig. 1. Infant mortality rates (deaths per 1000 births) at three different spatial scales and their approximate counterparts using census administrative areas (legend for maps on the right is not shown)

The ability to control the spatial basis of support for cancer rates is the key idea that geographic information systems bring to the task of providing decision support for cancer prevention and control. A key question we ask is at what geographic scale do significant differences in cancer incidence rates or other measures of cancer exist in any region of interest? A reason for asking this question is so that we can decide the scale at which interventions should be planned. Logical though this question may appear, it has not been the question that has driven the rather large literature of spatial analysis of cancer. Traditionally, cancer maps were based on predefined political or administrative units for which cancer data was collected. Starting with regions already defined we made maps and then asked "do we see a pattern." Such a strategy pre-supposes that spatial variations that occur within the regions mapped do not exist or, if present, are not relevant or important. With GIS, however, we start with geocoded data—at the level of points or small areas—and then we ask "at what geographic scale do we want to view this pattern?" Thus, it is the much smaller literature of spatial analysis of cancer based on data manipulated in a GIS that is the literature most relevant to cancer control and prevention. Cancer maps for this purpose employ density estimation methods. Unlike traditional cancer maps that show cancer statistics based on spatial units of different sizes, shapes and populations that conceal scale dependent patterns, density estimation techniques are designed to control the spatial basis of support for the spatial pattern of any statistic of interest. These are made possible by developments in the availability of geospatial data, geocoding techniques, and methods of spatial analysis that allow the opportunity to control the size, shapes and population characteristics for the spatial units for which statistics are computed.

1.4 Mapping cancer burdens

The first measure of the cancer burden on a population is the rate of incidence of any particular cancer type adjusted for age and sex of the local population. The first choice to be made is between direct and indirect rate adjustment methods. Direct adjustment of rates is made when rates are to be compared from one area to another to note the rate burden on the population. In such a situation the question being asked is the hypothetical question "if the age-sex structure of the local population was the same as a standard population, what would the overall cancer incidence rate be? These rates are made by multiplying locally observed age-sex defined cancer rates by a common set of weights that sum to one that describe national population characteristics, (see Pickle and White 1995). Indirect adjustment of rates are made when the question being asked is "if the local population were to have cancer incidences at the same rates as a standard population, how much more or less does cancer occur there than in the standard population." Indirectly adjusted rates are best used when resources are to be allocated to areas based on the impact of the rates on the population of the local area—see Kleinman 1977. The second choice of cancer burden is about the proportion of diagnosed cancer cases that are late stage at the time of their first diagnosis. This can be measured as the proportion of incidences observed in a population that are late stage, or, can be measured as the number in a population adjusted for its age and sex characteristics. The third choice is mortality rates. Illustrations of the different kinds of maps of these three cancer burdens for the Iowa population between 1998 and 2002 can be seen at Beyer et al. 2006. All maps are indirectly age-gender adjusted using national rates of cancer with the rates defined as actual observed number of cancers in the spatial filter area divided by the number expected given the demographic characteristics of people in the filter area. Rates defined in this way reflect the demographic characteristics of the local area. Statistically they are more robust than directly age-gender adjusted rates because they are made by multiplying national rates that are stable by populations in the filter areas which are also stable. The geographic detail in the indirectly adjusted rate maps is far superior to the geographic detail possible in directly age-sex adjusted maps.

I illustrate the control of scale with design of a map of late stage colorectal cancer rates in Iowa for the period 1993 through 1997. The approximate population of Iowa in 2000 was 2,800,000. The number of new incidences of colorectal cancer in Iowa for a four year period was 8,403 cases. All were geocoded either to their street address, or, in a few cases where the street address could not be matched to the geographic base files to the centroid of their Zip code; there are 940 Zip code areas in Iowa. Using a regular grid of four miles, we applied the "sliding window" method of Weinstock (1981) for estimating the late-stage rate at each node of this regular grid. For the area surrounding each node on this grid, the rate of late stage diagnosis is the ratio of the number of late stage colorectal cancers to the total number of colorectal cancers within the filter area (or kernel). In Figure 2 we illustrate the grid points from which the late stage colorectal cancer rates were constructed. On the right, the rates are illustrated as average values for the closest eight neighbors to each grid point, using an inverse distance weighting algorithm. In Figure 3 we change the scale of the patterns by using progressively larger spatial filters from ten miles radius to fifteen miles radius. In this illustration, we are mapping the rate with which women diagnosed with early stage breast cancer selected breast conserving surgery (lumpectomy with radiation) rather than the more radical surgery—mastectomy. As is to be expected on all disease rate maps, as the geographic scale of the map decreases (larger spatial filters), details in the pattern—many of which are spurious because the rates are based on small numbers—drop out and a more persistent regional pattern emerges which is best seen on the fifteen mile filter map. The named places on these maps had radiation facilities at the time of this data—early 1990s.

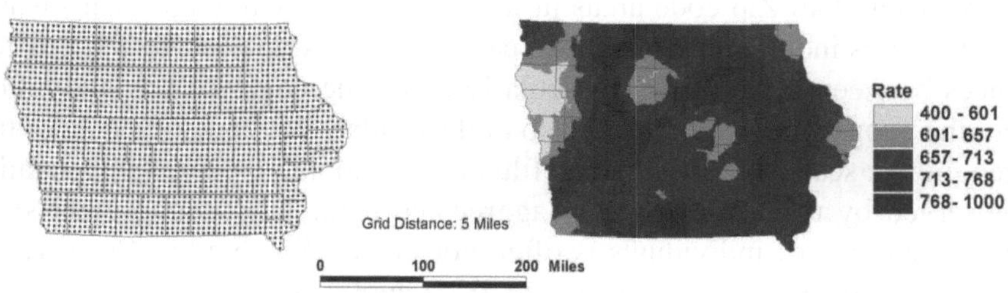

Fig. 2. Late stage colorectal cancer (number late stage per thousand cases of colorectal cancer diagnosed) interpolated from computed values on the regular grid (left)

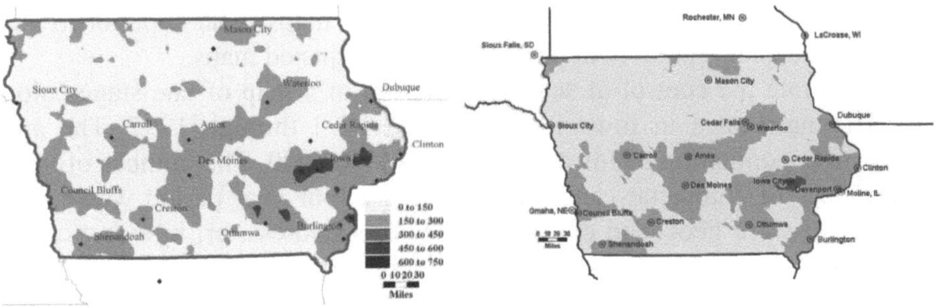

Fig. 3. Number of women selecting lumpectomy with radiation per 1000 cases of localized breast cancer, Iowa, 1991-1996; map on the left used 10 mile spatial filter; map on the right used 15 mile filter

The maps illustrate the tendency for women who live far from radiation facilities to not choose this recommended surgical therapy over the traditional more radical surgery of mastectomy. Recent research confirms that this tendency is a national phenomenon (Nattinger et al. 2001; Schroen et al. 2005). The critical choice in such spatial filtering of disease data are selection of the size of the grid and the size of the filter (Silverman 1986). The grid size is the less important choice since providing the grid is detailed enough geographically to provide the level of resolution desired in the output, further detail in the grid will add no further value to the map. Changing the size of the spatial filter, as illustrated in Figure 3, will affect the pattern because the differences in rates that typically occur within the size of the filter will be averaged or smoothed and some of the variability in the geographic pattern will disappear.

The geographic detail of a disease density map does depend on the level of spatial aggregation of the data used. Figure 4 illustrates late-stage colorectal cancer rates for the case (left map) where input data consists of approximately 940 Zip code areas in Iowa compared with (right map) where input data is individually geocoded cancer cases. Note that there are differences between these maps, particularly along the edges of the study area; but the geographic patterns are also quite similar. We conclude that, at this geographic scale—15 mile radius filters—considerable geographic detail is preserved by using the spatially aggregated data. This is important since geocoded data of individuals is often not made available by cancer registries in North America to researchers or to public health personnel because of privacy laws and commitments to maintaining the confidentiality of data records (CDC 2003; Olson et al. 2006). The improved geographic detail may also be compared with Figure 5 where area-based disease maps are based on the same data aggregated by county.

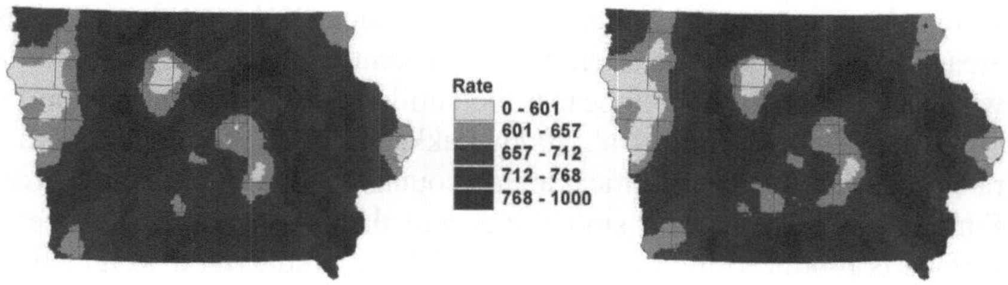

Fig. 4. Comparison of spatially filtered maps (15 mile filters) using geocoded cancer data at two different levels of spatial resolution. Rates of late-stage colorectal cancer at first diagnosis 1993-1997. The map on the left is made from spatially aggregated data which used Zipcode centroids as geocodes. The map on the right used address-matched geocodes. The same cancer incidence data is used on both maps.

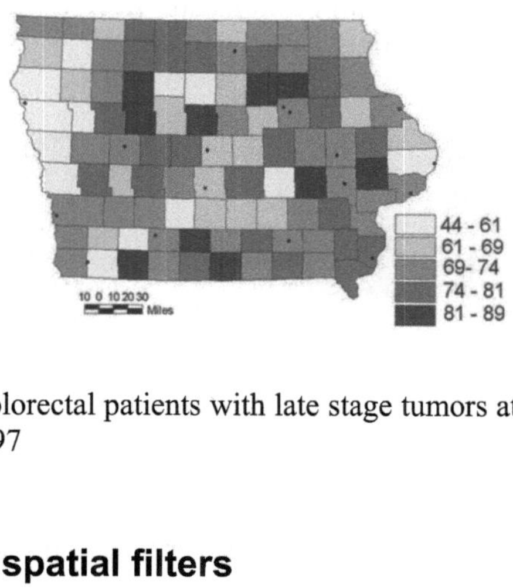

Fig. 5. Percent of colorectal patients with late stage tumors at time of first diagnosis, Iowa, 1993 - 1997

1.5 Adaptive spatial filters

Further geographic detail may be achieved by adapting the size of the spatial filter to the density of the disease data.

In Figure 6 the map on the left aggregates the cancer cases in order of their distance from the grid point until at least 100 cases are found. The map on the right of this figure shows percent late-stage colorectal cancer rates based on a 24 mile filter. The spatially adaptive filter provides more geographic detail in areas of high population density where the numbers of cancer cases within any given size spatial filter area is large enough to support a reliable estimate of the late-stage rate—Tiwari and Rushton 2004; Talbot et al. 2000. The spatial detail provided by such maps should not be confused with the apparent geographic detail on maps that are smoothed using rates for administrative or political entities (Kafadar 1996). Such maps use spatial smoothing functions based on centroids of areas. Examples can be seen in two recently published cancer atlases which superficially may appear to be similar to the mapping method proposed here—Tyczynski et al. 2006; Pukkala et al. 1987. In these atlases, rates are computed for political areas (counties in Ohio; municipal areas in Finland)—first-level data smoothing--and then the smoothed cancer rate surface is produced by a floating spatial filter producing a weighted average of the rates in surrounding counties—second-level data smoothing. This double smoothing of data and then rates, we believe, should be avoided. In kernel density estimation the data for numerator and denominator are collected for the spatially adaptive area and then the rate is computed and attributed to the grid point from which the kernel is measured. This method for controlling the change of support (see Haining 2003, p 129) is theoretically more valid than the gross spatial smoothing functions so commonly used. Spatial interpolations are made only locally; that is, between closely spaced grid points.

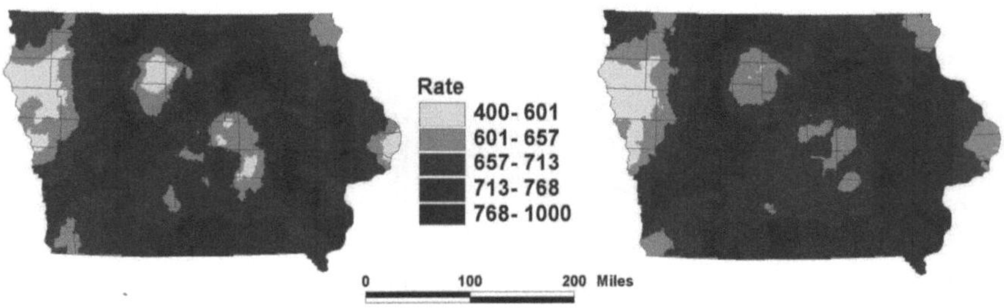

Fig. 6. Adaptive spatial filter--left map uses closest 100 cases to define the filter area from each grid point on a three mile grid; right uses a 24 mile filter area. Number of late stage colorectal cancer cases per thousand cases at first diagnosis, Iowa, 1993 – 1997

1.6 Adjusting for rate variability due to small numbers

The issue of reliability of rates is important. With traditional spatial density maps that use fixed size spatial filters, some local rates are based on a large amount of information while other local rates are based on little information. A Monte Carlo procedure can be used to evaluate the statistical significance of rates observed at any grid point. For this procedure we use random re-labeling of the known cancer locations so that the total number of late-stage cases in the study area is equal to the observed number of such cases. Thus the null hypothesis being tested is that the rate of late-stage colorectal cancer for this time period was uniform across the state of Iowa. We computed 1,000 simulated maps of late-stage colorectal cancer based on the probability that each colorectal cancer incident case is late-stage according to the statewide rate. For each of these simulated maps we compute the rate of late stage at each grid location. We then compute the proportion of the simulated rate maps at each grid location that are smaller than the observed late-stage rate. This is known as a p-value (probability value) map. Figure 7 illustrates these proportions for the colorectal cancer map shown in Figure 6. Because this method of measuring reliability involves multiple tests of the hypothesis that the observed rate is greater than the rate of the null hypothesis, these proportions are not equivalent to conventional significance rates—for a discussion of true maps of significance, see Kulldorff 1997 and 1998. This is a well-known feature of the work of Openshaw et al. (1987) where a similar approach was first used. Further details of this test procedure are provided in Rushton et al. 1996.

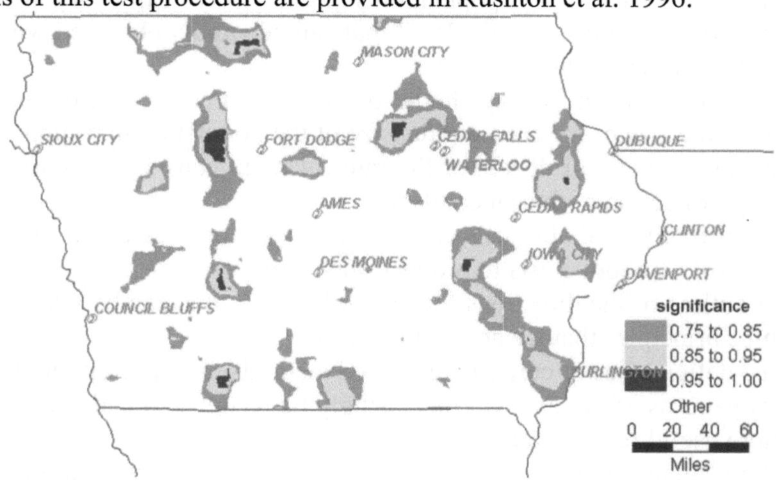

Fig. 7. P-value map showing results of Monte Carlo test of hypothesis that no areas are significantly different from the state wide rate of late-stage colorectal cancer, Iowa, 1993-1997

2 Conclusions

Different measures of the cancer burden can be mapped at reasonably local geographic scales. These maps can be used by both health professionals and the public to guide policy making, decision making, and action. The spatial basis of support for cancer statistics that are mapped needs more research and experimentation. Spatial density estimation techniques have been comparatively neglected in favor of inferior spatial analysis approaches that have focused on given, often inappropriate, spatial units. The drawbacks of current mapping approaches are well-known. Rates for areas with different population sizes differ in their reliability and many statistical methods and spatial smoothing methods are used to compensate and adjust for these problems. These mapping approaches do not convey clearly the geography of the cancer burden to local communities in a form that satisfies the needs of the public. Better methods exist but they have not been used with currently available geo-spatial population data and geocoded cancer registry data largely because software to make such maps is not available for general registry use. There are three essential properties of more useful mapping methods:

1. Rates mapped should be based on control of the population basis that supports them;
2. The spatial basis of this population support will typically vary in size of area so that the geographic detail that can be validly observed will typically vary across the map;
3. The user of a cancer map should be able, for any location on the map, know the size of the area and the size of the population that supports the rate as well as full details of the rates mapped consistent with full privacy protection of the cancer data.

No currently available cancer map has these three essential properties and no software tool exists to produce such a map.

An outline of the directions for research in this area can be made, based on three principles that we accept to be true:

- The deficiencies of current area-based methods for representing the spatial patterns of disease will increasingly be recognized and demands for more useful representations will grow;
- The availability of finely geocoded disease data will grow although access to such data will be increasingly tightly controlled through data sharing agreements and legal regulations, (see Rushton et al. 2006);
- The availability of demographic data for very small areas will grow as modern censuses tabulate data for flexible, GIS controlled, areas and as algorithms are developed for more intelligent disaggregating of demographic data to custom-defined areas, (see Cai et al. 2006; Mennis 2003; Mugglin et al. 2000).

Acknowledgments

I thank Chetan Tiwari for making Figures 2, 4 and 6.

References

[1] Abed J, Reilley B, Butler MO, Kean T, Wong F, Hohman K (2000) Developing a framework for comprehensive cancer prevention and control in the United States: an initiative of the Centers for Disease Control and Prevention. Jn Public Health Management Practice 6(2):67-78

[2] Armstrong B (1992) The role of the cancer registry in cancer control. Cancer Causes and Control 3:569-579

[3] Beyer K, Chen Z, Escamilla V, Rushton G (2006) Iowa Consortium for Comprehensive Cancer Control Cancer Maps Site. Available at http://www.uiowa.edu/~gishlth/ICCCCMaps/

[4] Bithell JF (1990) An application of density estimation to geographical epidemiology. Statistics in Medicine 9:691-701

[5] Cai Q, Rushton G, Bhaduri B, Bright E, Coleman P (2006) Estimating small-area populations by age and sex using spatial interpolation and statistical inference methods. Transactions in GIS 10:577-598

[6] Centers for Disease Control and Prevention (2003) HIPAA privacy rule and public health: guidance from CDC and the US Department of Health and Human Services. Morbidity Mortality Weekly Report 52:suppl.1-20

[7] Devesa SS, Grauman DJ, Blot WJ, Pennello GA, Hoover RN, Fraumeni, Jr. JF (1999) Atlas of Cancer Mortality in the United States: 1950-94. NIH Publication No. 99-4564

[8] Haining RP (2003) Spatial Data Analysis: Theory and Practice. Cambridge University Press

[9] Kafadar K (1996) Smoothing geographical data, particularly rates of disease. Statistics in Medicine 15:2539-2560

[10] Kleinman JC (1977) Age-adjusted mortality indices for small areas: applications to health planning. American Jn. of Public Health 67(9):834-840

[11] Kulldorff M (1997) A spatial scan statistic. Communications in Statistics—Theory and Methods 26:1481-96

[12] Kulldorff M (1998) Statistical methods for spatial epidemiology: tests for randomness. In: Gatrell AC, Loytonen M (eds) GIS and Health, Taylor & Francis, Philadelphia, pp 49-62

[13] Mennis J (2003) Generating surface models of population using dasymetric mapping. Professional Geographer 55:31-42

[14] Mugglin AS, Carlin BP, Gelfand AE (2000) Fully model-based approaches for spatially misaligned data. Jn. of the American Statistical Association 95:877-887

[15] Nattinger AB, Kneusel RT, Hoffmann RG, Gilligan MA (2001) Relationship of distance from a radiotherapy facility and initial breast cancer treatment. Jn of the National Cancer Institute 93:1344-1346

[16] Olson KL, Grannis SJ, Mandl KD (2006) Privacy protection versus cluster detection in spatial epidemiology. American Jn. of Public Health 96:2002-2008

[17]Openshaw S, Charlton M, Wymer C, Craft AW (1987) A Mark I geographical analysis machine for the automated analysis of point data sets. International Jn. of Geographic Information Systems 1:335-358
[18]Pickle LW, Mungiole M, Jones GK, White AA (1996) Atlas of United States Mortality. Hyattsville, Maryland: National Centre for Health Statistics
[19]Pickle LW and White AA (1995) Effects of the choice of age-adjustment method on maps of death rates. Statistics in Medicine 14:615-627
[20]Pukkala E, Gustavsson N, Teppo L (1987) Atlas of Cancer Incidence in Finland 1953-1982. Vol. 37. Helsinki: Cancer Society of Finland
[21]Rushton G and Lolonis P (1996) Exploratory spatial analysis of birth defect rates in an urban population. Statistics in Medicine 15:717-726
[22]Rushton G, Krishnamurthy R, Krishnamurti D, Lolonis P, Song H (1996) The spatial relationship between infant mortality and birth defect rates in a U.S. City. Statistics in Medicine 15:1907-1919
[23]Schroen AT, Brenin DR, Kelly MD, Knaus WA, Slingluff Jr CL (2005) Impact of patient distance to radiation therapy on mastectomy use in early-stage breast cancer patients. Jn of Clinical Oncology 23:7074-7080
[24]Silverman BW (1986) Density estimation for statistics and data analysis. Boca Raton, FL, Chapman & Hall/CRC
[25]Talbot TO, Kulldorff M, Forand SP, Haley VB (2000) Evaluation of spatial filters to create smoothed maps of heath data. Statistics in Medicine 19:2399-2408
[26]Tiwari C and Rushton G (2004) Using spatially adaptive filters to map late stage colorectal cancer incidence in Iowa. In: Fisher P (ed) Developments in spatial data handling. Springer-Verlag, pp 665-676
[27]Tyczynski JE, Pasanen K, Berkel HJ, Pukkala E (2006) Atlas of Cancer in Ohio: Incidence & Mortality. The Cancer Prevention Institute, Columbus, Ohio
[28]Weinstock MA (1981) A generalized scan statistic test for the detection of clusters. International Journal of Epidemiology 10:289-293

Environmental Risk Factor Diagnosis for Epidemics

Jin-feng Wang

State Key Laboratory of Resources and Environmental Information System, Institute of Geographical Sciences and Nature Resources Research, Chinese Academy of Sciences

Abstract: There is evidence to suggest that the rapidly changing physical environment and modified human behaviors have disrupted the long-term established equilibrium of the chemical composition between human and the Earth environment. We have noticed that environmentally related endemic is increasingly persistent in poorer areas and occuring in rapidly developing regions. This chapter describes two models developed respectively to diagnose the risk of environmentally related diseases and to simulate the spatio-temporal spread of communicable diseases. In the first model, we used birth defects to show the diagnosis of an endemic by (i) detecting risk areas, (ii) identifying risk factors, and (iii) discriminating interaction between these risk factors. Here, a spatial unit is considered a pan within which multiple environmental factors are combined to exert impacts on the human which may lead to either positive or negative health consequences. We were able to show that a diagnosis of environmental risks to population health discloses the locations at risks and the potential contribution of environment factors to the disease. In the second case, we used SARS to show the modeling of a communicable disease by (i) inversing epidemic parameters, (ii) recognizing spatial exposure, (iii) detecting determinants of spread, and (iv) simulating epidemic scenarios under various environmental and control strategies. We were able to demonstrate spatial and temporal scenarios of the disease through the modeling of communicable epidemic spread.

Keywords: environmentally related diseases, spatio-temporal simulation, spatio-temporal modeling

1 Introduction

The modern society is characterized by rapidly changing physical environment and modified human behaviors which disrupt the long-term established equilibrium of the chemical composition between human and the earth environment. Increasingly, we have noticed that environmentally related endemic is persistent in poor areas and occurs in rapidly developing regions.

The causal factors and determinants of a disease are critical in its control and intervention. These factors could be in different levels, from micro gene, physiological, chemical or biological abnormality, to the macro media or geographical environment. Such factors at different levels could exist in a cause-effect chain or separately and independently impact the human bodies and causing diseases.

A GIS coupled with spatial analysis and spatial statistics offers powerful tools in exploring macro patterns and factors. The macro-level examination could suggest proxies on the visible surface of some obscured micro agents along the cause-effect chain to uncover the real and direct causes of a disease. Spatial analysis tools are now available to explore environmentally related diseases. The causal factors X could be investigated through cases, or a response variable y in the mathematical nomenclature, such as spatial pattern alignment between cases y and the proposed causes X; spatial ANOVA of y and x; and time series of X. This chapter looks at our efforts in employing spatial analysis tools in diagnosing environmental risk factors for diseases.

2 Inversion Epidemic Parameters

We started by exploring the inversion epidemic model which is stated simply as

$$\Theta = g^{-1}(Y) \qquad \text{Eqn. (1)}$$

where Y denotes reported cases of infections, Θ stands for epidemic parameters, g is a mechanistic equation of the variable Y, and -1 denotes an inverse transformation. The epidemic parameters reflect the essential features of an epidemic which correspond either to a unique cause or is a complete consequence of several factors.

Two approaches can be employed to derive the parameters: (i) a field survey which needs a huge amount of data collection work, and (ii) a model reversion. If the mechanistic model of a disease is known, then only a few cases must have the parameters reverted, because we know that

$$Y = \text{mechanism} \otimes \text{epidemic parameters} \qquad \text{Eqn. (2)}$$

where ⊗ denotes a combination in a broad sense between components on both sides of the symbol. When the first two items in Eqn (2) are known, the epidemic parameters can then be estimated.

We used the 2003 SARS data of Beijing to illustrate the philosophy (Wang et al. 2006). A communicable disease spreads in time in a mechanism described by a time varying parameter following the SEIR model:

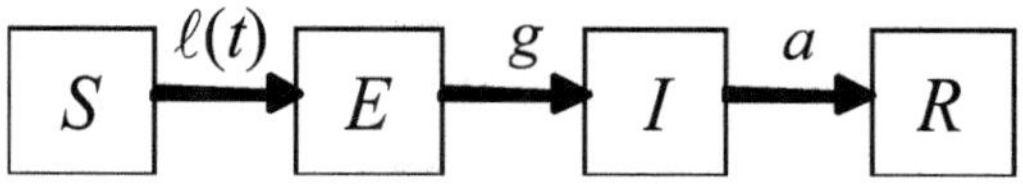

where $E(t)$, $I(t)$ and $R(t)$ are respectively the number of Exposed, Infectious and Removed individuals at time t. S denotes the population Susceptible to the disease. $I(t)$, the average number of contacts per infectious person, depends on time because the control effort changes over time. g is the rate at which the exposed (latent) individuals become infectious while a is the rate at which the infectious individuals are removed (recovered or isolated). The basic reproduction number for this model is given by $R_0 = l(0)/a \approx (b+c)/a$ and the eventual reproduction number is approximated by $R_0 \approx b/a$.

We fitted the model to the case incidence data of Beijing over the period between 19 April and 21 June in 2003 to obtain these parameter estimates: a = 0.252, b = 0.008, c = 0.588, d = 0.368, e = 54 and g = 0.200. Figure 1 shows a fitted curve for the number of infected individuals and a fitted curve for the transmission rate showing a very rapid decline over the period between 20 April and 30 April. The average incubation period was 1/g or about 5 days and the average infection period was 1/a or about 4 days. Our estimate of the basic reproduction number was 2.37. The eventual reproduction number, achieved at around 11 June, was found to be 0.1, indicating a dramatic reduction in the reproduction number. The total size or cases of the epidemic for estimating the epidemic parameters using the model was 2522.

The difference between the curve of the estimated infection rate and those of similar diseases confirmed that the model can disclose, to a certain extent, the strength of intervention if there was no abrupt change of other factors during the epidemic.

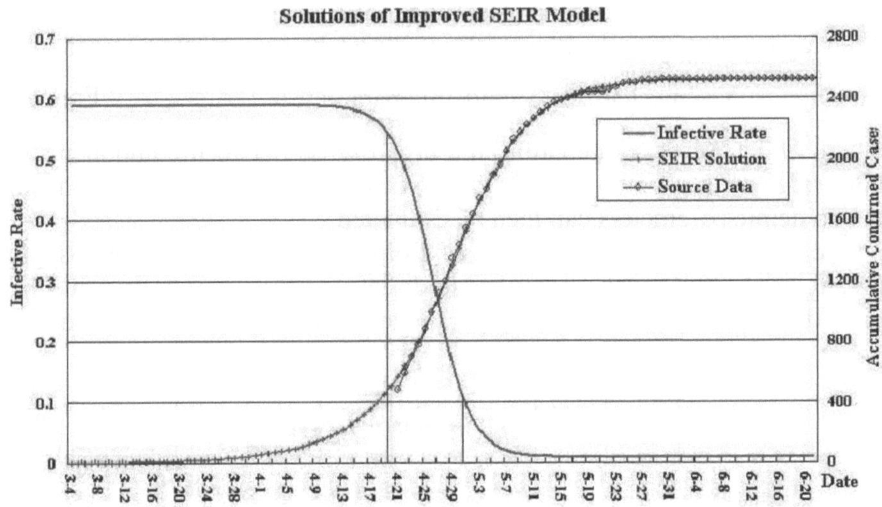

Fig. 1. The Beijing epidemic and its control over time

Following a similar argument for Eqn. (2), we regarded that

Estimated Y = Observed Y + Residual Eqn. (3)

Then, the residual is actually a factor which has not been included in the mechanism model.

3 Pattern Alignment

We also employed the spatial pattern alignment model to explore environmentally related diseases with some degree of success. The model is simply stated as

Pattern Y = Pattern X Eqn. (4)

where Y denotes cases of an infection and X the factors.

Birth defects, defined as "any anomaly, functional or structural, that is present in infancy or later in life (ICBDMS)", are a major cause of infant mortality and a leading cause of disability in China. The left side of Figure 2 illustrates the neural tube birth defect (NTD) prevalence in China; the right side of Figure 2 shows the NTD in Heshun County of the Shangxi province, which is the location of our pilot study.

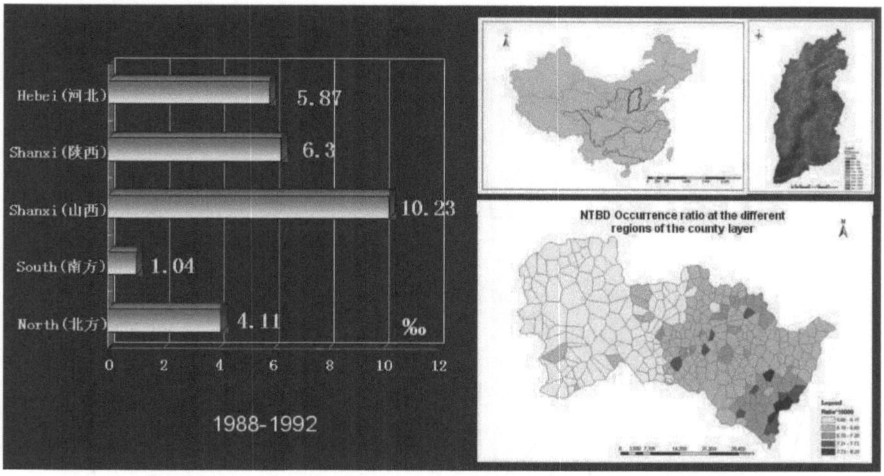

Fig. 2. Neural tube birth defect (NTD) prevalence in China

NTD is believed to be caused by a multitude of factors including hereditary, crude and artificially polluted environment, nutritional deficiency, and social, economic and behavioral factors (Figure 3). However, the risk factors associated with heredity and/or environment are very difficult to single out from our analysis. An exhaustive survey of each of the factors in the study area is possible but too expensive and time consuming to undertake.

We used the pattern alignment method to justify roles of the geological environment and the genetic factor in NTD within the study area (Wu et al. 2004). The NTD ratio was calculated according to birth defect registers from the hospital records and field investigations in villages over a four year period in 1998-2001. The ratio was adjusted by the Bayesian model to reduce variation in the records of a small probability event by borrowing strengths of its neighbors (Haining 2002). The Getis G* statistics (Getis and Ord 1992) was used to detect spatial hotspots of the ratios in different distance scales. Two typical clustering phenomena were found present in the study area.

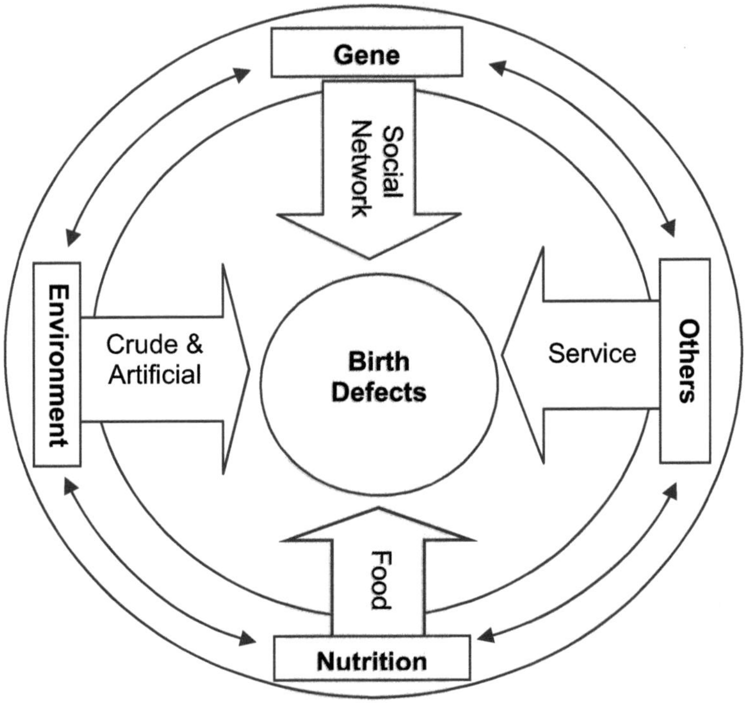

Fig. 3. Causes of birth defects

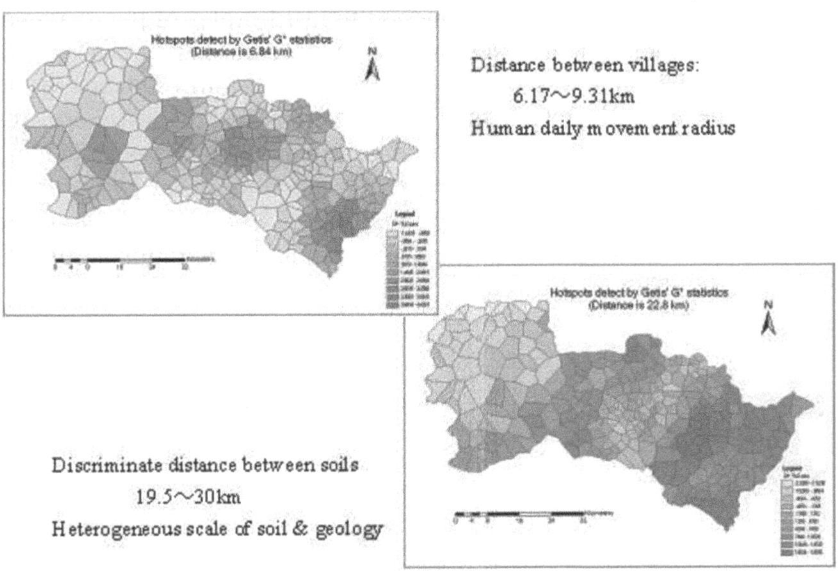

Fig. 4. Two scales of hotspots of NTD prevalence

The upper left display of Figure 4 unveiled spatial clusters at a distance of around 6.5 km, which corresponded with the average distance separation of 6.31-9.17 km among villages in the Heshun County. This average distance of social contact indicated very little mixing of inhabitants between villages further apart which would infer that hereditary might have a role in inducing NTD within the study area.

A macro belt pattern emerged when the scanning radius was increased to 19-30 km, as seen in the lower right display of Figure 4. This pattern matches almost perfectly with the geological and soil patterns of the study area shown in Figure 5. Accordingly, we could infer that the geological circumstance might also be a risk factor to NTD occurrence within the study area.

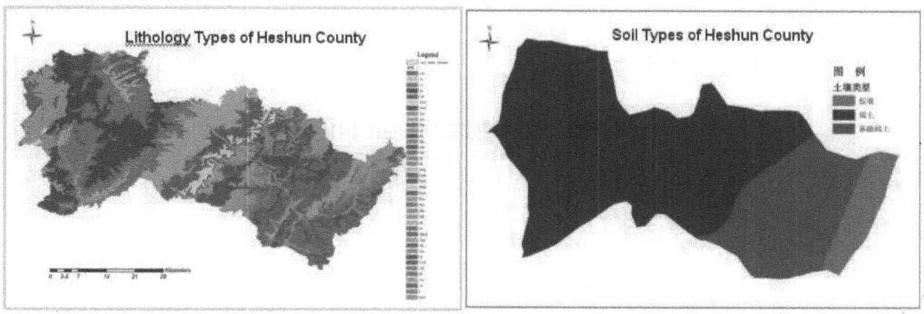

Fig. 5. Lithozone and soil distribution in Heshun County

The results of our spatial pattern alignment exercise have provided clues that NTD was likely an environmentally related disease. Such findings from the two spatial scales could be used to suggest further actions be taken, such as the conduct of more physical, chemical and even molecular laboratory tests.

4 Spatial Regression

To explore further the relationship between NTD and the geological structures of Heshun County (Li et al. 2006), we tried a spatial regression model as defined below:

$$Y = f(X) \qquad \text{Eqn. (5)}$$

where Y denotes the response variable; X the causal variable; and f a statistical function between Y and X. f could be a spatial linear regression function such as SAR, MA and CAR (Anselin 1988; Haining 2003); or a nonlinear function such as neural network, genetic algorithm and Bayesian network; or ANOVA; or just a scatter plot of Y and X. A significant statistical function *f* would suggest that the corresponding X is the most probable causal factor for the disease under examination.

The geological structure selected for the spatial regression was the locations of fault lines in Heshun County (Li et al. 2006). The study area was divided into eight zones of buffer distance from the geological fault lines: 0–2, 2–4, 4–6, 6–8, 8–10, 10–12, 12–14, and 14–16km (Figure 6). The NTD ratios within each of the eight buffer zones were computed and the values graphed against the buffer distances as shown in Figure 6. The graph discloses that the occurrence ratio of NTD birth defects was the highest in regions at about 4 km from a fault line and the reading decreases as the buffer distance increases. At the macro spatial scale, the geological background showed a significant correlation with the risk of birth defects and that people residing near the fault zones had a higher risk of having babies with birth defects.

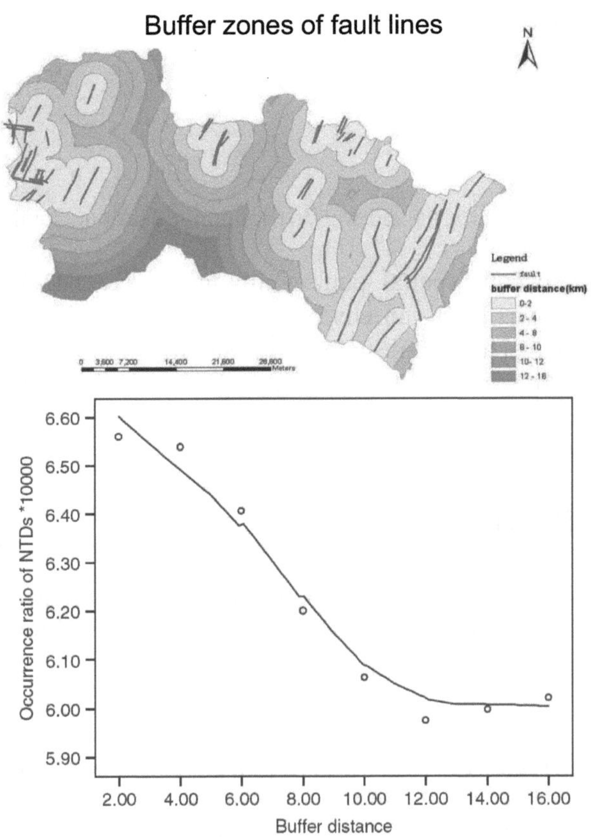

Fig. 6. Buffer zones of fault lines and their correlation with NTD ratio

A possible interpretation to the phenomena projected by Figure 6 can be deduced from other research observations (Trique 1999; Parker and Craft 1996), such as a higher concentration of radon in the soil, water, and air near a fault zone. The radiation emitted by radon and those of its daughter products comprises predominantly of high linear energy transfer (LET) alpha radiation. Studies have indicated that the relative biological effectiveness (RBE) of the LET alpha radiation as emitted by radon and its daughter products is 20-fold higher than those of the X-ray and gamma radiation. A dose of the LET alpha radiation will put a fetus in the uterus in great risks of developmental damages.

5 Time Varying Factor Detection

The disease prevalence over time can be examined with the following model,

$$Z\{BW(Y) \mid Net(X)\} \sim N \qquad \text{Eqn. (6)}$$

where Y denotes the number of newly reported infection; X the proposed factors of Y; and N a normal distribution. In this model, two districts are considered connected if a linkage is the proposed channel for the transmission of Y; where a linkage can be a real geometric neighbor or a consecutive district in a ranked hierarchy according to population density or other measures. For each day within the study period, a district was coded black (B) if a disease case was reported on that day; otherwise it was coded white (W). Each join in the network would link two B districts or two W districts or a pair of B and W districts. These joins were labeled as BB, WW and BW, respectively. The observed number of BW joins was compared against the expected value, and a standard normal deviation (z-scores) was used to test its significance (Haggett 1976). A high negative value of the z-statistics would indicate evidence of a clustering of cases in the network and a high positive value evidence of scattering.

We investigated associations between environmental factors and SARS by considering seven possible networks for the spread of infection among the districts in Beijing in spring of 2003 (Meng et al. 2005). The networks were assessed against the data using the BW join-count test. The seven networks are listed below:

- N1: Local transmission: two districts were connected if they shared a common geographic boundary
- N2: Nearest district: a district was connected to its nearest neighbor as determined by distances between their polygon centroids

- N3: Population size: districts were ranked according to population size and consecutive districts in this hierarchy were connected
- N4: Population density: same as N3 but ranked by population density
- N5: Number of doctors: same as N3 but ranked by the number of doctors in a district
- N6: Number of hospitals: same as N3 but ranked by the number of hospitals in a district
- N7: Urban-rural: Eight districts were designated as urban while the remainder as rural. A rural and urban pair was connected if (i) they shared a boundary, (ii) the urban district could be reached from the rural district by passing through just one other rural district, or (iii) the rural district could be reached from the urban district by passing through just one other urban district.

For each network, we calculated the BW join-count statistics for each day, and plotted the changes of this statistics over time.

Figure 7 shows associations between various environmental factors and the spread of SARS infection between 27 April and 25 May 2003. The diagrams are annotated a horizontal line indicating the threshold significance value of the z-statistics. Figure 7(a) shows the number of cases and the number of infected districts over time. There was a clear indication or strong evidence in Figures 7(b) and 7(c) of transmission between the neighboring districts towards the end of April. This local transmission continued into the first week of May but was not significant thereafter. Between 13th and 19th of May, there was clear evidence of spread between the urban and rural areas, again a reflection of the outbreaks in the Tongzhou district at about that time. The remaining factors showed sporadic associations with the spread of SARS, suggesting a relationship between diffusion of infection and both the number of doctors and the population density.

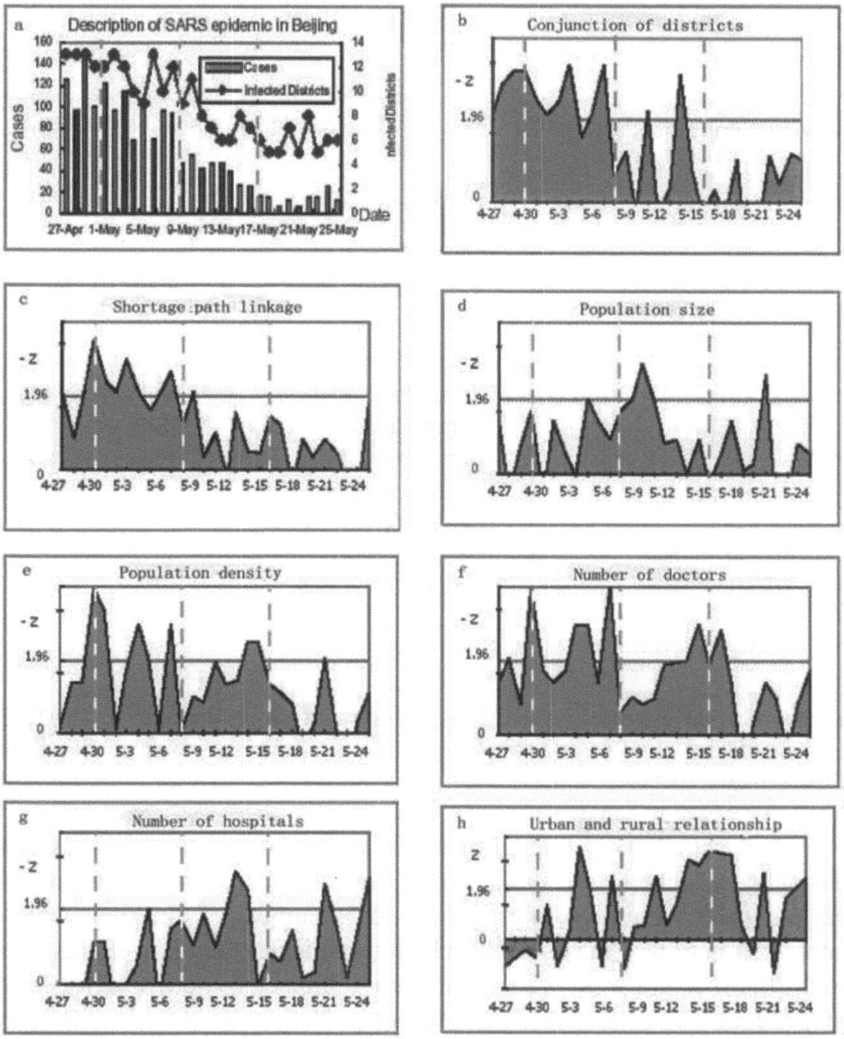

Fig. 7. Relationship between incidence of new cases and various factors of spatial spread

6 Conclusion and Discussion

The approaches to studying both environmental and communicable diseases have been documented in this chapter. Four approaches have been employed to explore relevant factors of these epidemics. Figure 8 summarizes an integrative framework and prospective of these approaches to better understand factors leading to specific patterns of spatial spread of diseases.

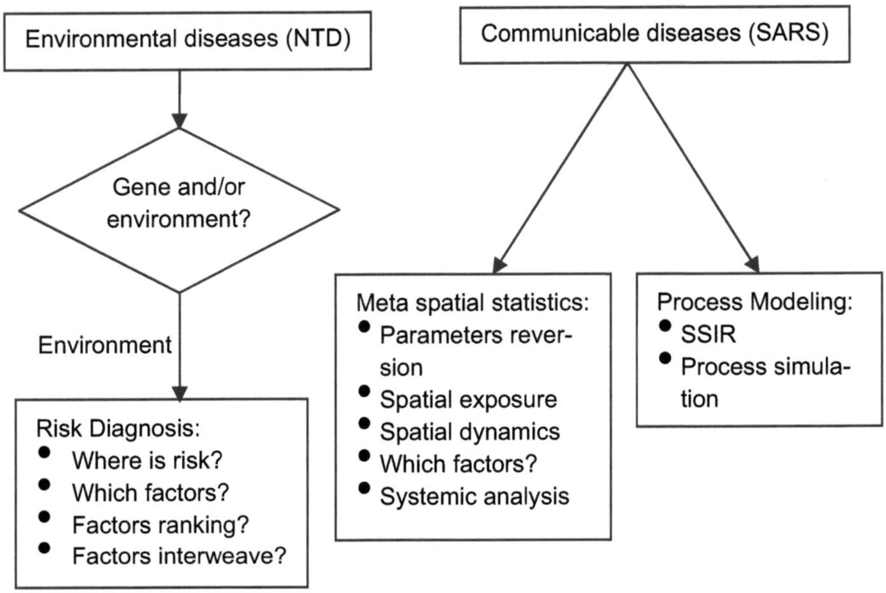

Fig. 8. Spatial statistics, systematic analysis and process modeling (NTD denotes neural tube defect; SARS denotes Severe Acute Respiratory Syndrome; SSIR is Spatial SIR)

For environmentally related diseases, we attempted firstly to discriminate between genetic and environmental roles. The Getis G* statistics of varying scales were used toward this end. A number of questions were of concern: Where was the risk? Which environmental factors were responsible for the risk and what were their relative impacts? Did the factors work independently or collectively upon a human body to lead to diseases? Our approach considered the spatial unit as a platform within which multiple environmental factors were combined to exert impacts on the human to result in either positive or negative health consequences. Spatial sampling estimation, spatial regression and ANOVA were found suitable to addressing the above questions.

For communicable diseases, we highlighted the importance of epidemic parameters. These parameters were essential to recognize spatial exposure, detect determinants of spread, and simulate epidemic processes in space and time using scenarios under various environmental and control strategies. Spatial statistics and system modeling were successfully employed to handle these issues. The different spatial statistics could be joined by common items to produce a system fit for modeling both communicable and environmentally related diseases.

The invention of the microscope has helped humans discover the mechanisms of life in microscopic scales from cells to genes. The same analogy can be used in GIS which have made visible environmental factors related to life and linked these factors to a single or multiple groups of the population. By the same token, spatial analysis has helped investigations of relationships between human health and the proposed factors through the abilities to integrate many kinds of spatially related information.

The huge amounts of data from the micro genome to the macro digital earth have promoted the developments of bioinformatics and geoinformatics respectively. Pattern alignment and correlation, spatial prediction, hotspot detection are current topics of research interests in both bioinformatics and geoinformatics. It has been said that the two distinct disciplines have at least a 70% overlap in the mathematics employed, including the Bayesian inference, dynamic program, Markov chain, simulated annealing, genetic algorithm, probability likelihood, cluster, HMM (hind markov model), SVM (support vector machine), CA (Cella Automa), etc. (Baldi and Brunak 1998; Haining 2003). With findings accumulated from fundamental research and improvements made thus far, both tools are now employable in human decision making; for example, finding petrol deposits and searching for pathogenic genes, producing GIS products and inventing genetic pharmacy, and building government decision support systems and carrying out genetic therapies.

Besides sharing a common methodology and philosophy, bioinformatics and geoinformatics are influencing one another. For example, GIS is used in displaying and undertaking spatial analysis for genome studies (Dolan 2006) and in tracking global genetic change and global climate warming (Balanya 2006) in addition to evidence from glaciers and lake sedimentation. The micro mechanism of a macro phenomenon is also revealing; for instance, 99% of mouse genes have homologues in man or diverged from a common ancestor (http://www.evolutionpages.com/Mouse genome genes.htm#Homologues). The macro phenomena also provide clues for micro mechanisms; for instance, the two level spatial clusters of NTD prevalence in Heshun County suggested that both hereditary and geological factors had controlling influence of the disease in the area (Wu et al 2004); micro pathogenic agents were found to react under both physiological and natural environments; and the macro transmission of an epidemic could be modeled successfully (Nakaya et al. 2005; Xu et al. 2006). The molecular epidemics and genetic epidemics have emerged to explain macro phenomena from the micro mechanisms and the focus of investigation on life processes has moved from a single genome analysis to a more integrated protein function group. The above progresses made in bioinformatics and geoinformatics have suggested that a model can mechanically link micro and macro processes and an examination of the relevant factors would be necessary to impart a better understanding on the systems of life.

We highlighted three challenges in the use of spatial statistics in epidemics research: (1) small sample problems, (2) problems of large amounts of data versus large amounts of variables, and (3) problems in map comparison. Firstly, there is a high variation in the sample size of rare diseases or diseases with short term records, which often leads to biases in the confidence interval of estimation. Although such a variation can be reduced somewhat by the empirical Bayesian-adjusted technique which borrows strength from the neighbors or by inserting artificial data through stochastic simulations (Rushton and Lolonis 1996), we need much more reliable theories to handle the problem. Secondly, more variables are now available following better cooperation and coordination of the globally connected communities (Goodchild and Haining 2004) but disease incidence cases or samples have decreased because of better human controls. There is also the dilemma of wanting local data and analyzing data accurate to the individual level while having, at the same time, the ability to compile a global statistics. We need to explore diseases with a large amount of variables but having a few cases. This situation is contrary to the large number theory upon which modern statistics is based, thus presenting new challenges to existing statistical theories and models. Finally, spatial patterns of some diseases change with time, and comparison between disease case patterns and those of suspected factors should reduce the uncertainty of the hypotheses. But patterns and shapes are difficult to describe in natural languages, which form the prerequisite for artificial inference. There is here a need to explore more efficient tools for shape analyses.

Acknowledgement

This work was supported by the National Science Foundation under Grants #40471111 and #70571076, and by the 973 Project under Grant #2001CB5103.

Reference

[1] Balanya J, Oller J, Huey R, Gilchrist G, Serra L (2006) Global Genetic Change Tracks Global Climate Warming in Drosophila subobscura. *Science* 313(22): 1773-1775

[2] Baldi P and Brunak S (1998) *Bioinformatics - The machine learning approach.* MIT Press, Cambridge MA

[3] Dolan M, Holden C, Beard M, Bult C (2006) Genomes as geography: using GIS technology to build interactive genome feature maps. *BMC Bioinformatics* 7:416 (19 September 2006)

[4] Getis A and Ord JK (1992) The analysis of spatial association by use of distance statistics. *Geographical Analysis* 24:189-206

[5] Goodchild M and Haining R. (2004) GIS and spatial data analysis: Converging perspectives. *Papers Reg. Sci.* 83, 363-385

[6] Haggett P (1976) Hybridizing alternative models of an epidemic diffusion process. *Economic Geography* 52,136-146

[7] Haining R (2003) Spatial Data analysis, Theory and Practice. Cambridge University Press

[8] Li XH, Wang JF, Liao YL, Meng B, Zheng XY (2006) A geological analysis for the environmental cause of human birth defects based on GIS. *Toxicological & Environmental Chemistry* 88(3): 551–559

[9] Meng B, Wang JF, Liu JY, Wu JL, Zhong ES (2005) Understanding the spatial diffusion process of severe acute respiratory syndrome in Beijing. Public Health 119: 1080–1087

[10] Nakaya T, Nakase K, Osaka K (2005) Spatio-temporal modeling of the HIV epidemic in Japan based on the national HIV/AIDS surveillance. *Journal of Geographical Systems* 7: 313–336

[11] Parker L and Craft AW (1996) Radon and childhood cancers. *Eur. J. Cancer* 32:201–204

[12] Rushton G and Lolonis P (1996) Exploratory spatial analysis of birth defect rates in an urban population. *Statistics in Medicine* 15: 717-726

[13] Trique M (1999) Radon emanation and electric potential variations associated with transient deformation near reservoir lakes. *Nature* 399:137–141

[14] Wang JF, McMichael AJ, Meng B, Becker NG, Han WG, Glass K, Wu JW, Liu XH, Liu JY, Li XW (2006) Spatial dynamics of an epidemic of severe acute respiratory syndrome in an urban area. *Bulletin of the World Health Organization.* In press

[15] Wu JL, Wang JF, Meng B, Chen G, Pang LH, Song XM, Zhang KL, Zhang T, Zheng XY (2004) Exploratory spatial data analysis for the identification of risk factors to birth defects. *BMC Public Health* 4:23

[16] Xu B, Gong P, Seto E., Liang S, Yang Y, Wen S, Qiu D, Gu X, Spear R (2006) A spatial-temporal model for assessing the effects of intervillage connectivity in Schistosomiasis transmission. *Annals of the Association of American Geographers* 96(1): 31–46

A Study on Spatial Decision Support Systems for Epidemic Disease Prevention Based on ArcGIS

Kun Yang, Shung-yun Peng, Quan-li Xu and Yan-bo Cao

Faculty of Tourism and Geographic Science, Yunnan Normal University, Kunming, China.

Abstract: Having analyzed the current status and existing problems of Geographic Information Systems (GIS) applications in epidemiology, this chapter proposes a method to establish a spatial decision support system (SDSS) for the prevention of epidemic diseases by integrating the COM GIS, spatial database, gps, remote sensing, and communication technologies, as well as ASP and ActiveX software development technologies. One important issue in constructing the SDSS for epidemic disease prevention concerns the incorporation of epidemic spread models in a GIS. The chapter begins with a description of the capabilities of GIS in epidemic prevention. Some established models of an epidemic spread are studied to extract essential computational parameters. A technical schema is then proposed to integrate epidemic models using a GIS and relevant geospatial technologies. The GIS and modeling platforms share a common spatial database and the modeled results can be visualized spatially by desktop and Web clients. A complete solution for establishing the SDSS for epidemic disease prevention based on the model integrating methods and the ArcGIS software is suggested in this chapter. The proposed SDSS comprises several sub-systems: data acquisition, network communication, model integration, epidemic disease information spatial database, epidemic disease information query and statistical analysis, epidemic disease dynamic surveillance, epidemic disease information spatial analysis and decision support, as well as epidemic disease information publishing based on the Web GIS technology. The design process and sample VC and VB programming codes of the epidemic case precaution are used as an example to illustrate the basic principles and methods of the system development that integrates GIS functions with models of epidemic spread. A case study of AIDS in the Yunnan Province of China exemplifies the systems spatial analytical functions through its spatial database access and statistical analysis tools.

Keywords: epidemic disease spread models, model integration methods, epidemic spatial database, spatial decision support systems

1 Introduction

Epidemic diseases (such as SARS, AIDS and bird flu) are highly contagious and pose a threat to human life, hindering social and economic development and progress. Outbreaks and uncontrollable spread of an epidemic disease can cause serious public health problems of social and political significance. Spatial information systems containing spatial and temporal data on epidemic diseases and their application models can help a government and its public health institutions to realize disease monitoring and surveillance. Such a system has been known to uncover relations between a disease and its geographical environment, as well as to offer decision support in preventing an epidemic from spreading [11].

The first application of spatial epidemiology dates back to 1854 when John Snow succeeded in locating the origin of cholera in London by linking the disease with water pumps on a map at the local scale. Further development of the spatial information technologies in public health took place in the developed countries. These research developments have resulted in a range of tools and methodological approaches, including mathematical models of epidemic disease spread, integration of epidemic spread models in GIS, spatial and temporal epidemic analysis modeling, and preventive what-if scenarios. The scientific and technological achievements offer essential backgrounds and foundations for this research.

The unrelenting problems about the use of spatial information technologies in epidemiology are none other than insufficient data, deficient spatial and temporal modeling procedures and inadequate integration of epidemic models in GIS. Likewise, the problems of spatial epidemic disease databases and the integration of epidemic disease spread models in a GIS are the key issues to resolve in the construction of an SDSS for epidemic disease prevention. Using the ArcGIS software platform, a method is proposed here for the development of the SDSS.

2 Epidemic Spread Models and Integrated Applications in GIS

2.1 Roles of the GIS in Epidemic Disease Prevention

GIS is a technology to deal with spatial and temporal data. It offers visualization and spatial analysis tools to monitor the spread of an epidemic disease. It is suitable for the development of a disease tracking and prevention system given its spatial data acquisition and processing abilities, as well as its powerful spatial analysis functions.

The origin and subsequent spread of epidemic diseases have a close relation with time and geographic locations. If disease data are captured in space/location and time and they contain essential disease attributes, the spatial distribution and temporal characteristics of the disease spread may be monitored and visualized for probable intervention. With the availability of disease spread models, the contagious process may be dynamically simulated and visualized in two or three dimensional spatial scales. Consequently, high-risk population groups may be identified and visually located while the spatial distributional patterns and spreading behaviors of a disease may be uncovered. More effective prevention decisions may be made by the government and public health institutions through better allocation of medical resources by using the network analysis models of a GIS.

The use of GIS technologies in epidemic disease modeling and prevention will not only promote mutual developments in epidemiology and geographic information science, but will also promote the formation and development of spatial epidemiology, which has significant theoretical and practical values given an increased global concern over communicable diseases[13].

2.2 Mathematical Models on Epidemic Spread

Several popular models on epidemic disease spread include the SIS/SIR, Smallworld Network Dynamic and Cellular-Automata models, among which the SIS/SIR models are the most popular and mature epidemic models. Grassberger presented the SIR model to describe spreading behaviors of an epidemic over a network[3] and proposed that the network based SIR model may be analogous to percolation issues in the network[9]. His finding was later extended by Sander to deal with more general situations[4].

In the following equation, if β and γ are not the same at every node on the network but fit for $P_i(\beta)$ and $P_r(\gamma)$ distribution, Newman has proved that the network based SIR model is equal to a percolation issue[5] with the bond occupying rate of

$$T = 1 - \int_0^{\infty} P_i(\beta) P_r(\gamma) e^{-\beta/\gamma} d\beta d\gamma \qquad \text{............. Eqn. 1}$$

Therefore, the analytical solution of a network based SIR model may be obtained if the topological structure is given and the deterministic $P_i(\beta)$ and $P_r(\gamma)$ distribution can be confirmed.

However, the SIR model is not suitable for patients with diseases such as tuberculosis and gonorrhea because these patients may not be immune from the diseases and may become susceptible to the illnesses again. For these diseases, the SIS model described below is more appropriate:

$$\frac{ds}{dt} = -\beta is + \gamma i, \quad \frac{di}{dt} = \beta is - \gamma i \qquad \text{.................. Eqn. 2}$$

The SIS model is different from the SIR given that the network based SIS model may not obtain an accurate solution. Pastor-Satorras and Vespignani provided a very good approximate solution by using the mean field method[6][7].

There are many more epidemic spread models for various epidemic diseases besides the SIR and SIS models described above. For diseases with a limited immunity period, the SIR model is usually used to analyze the spreading behaviors. For diseases with latent periods, the concepts of latent population groups must be accounted for relevant estimations[2].

2.3 Integrating Epidemic Models in a GIS

The integration of application models in a GIS has been a research hot spot since 1980s. Broadly speaking, three technical aspects must be dealt within model integration. These include data management (i.e. acquisition, processing, storage and organization), model computing (i.e. application analysis), and model output (i.e. expressing simulated results in graphic or cartographic displays). Model integration in practice must consider specific application requirements because of different application settings. The COM based integration method has become a popular and feasible choice of model integration in SDSS development because of its flexibility, although the model base scheme is more ideal for model management and intelligent processing[3].

Looking at parameters in the equations of epidemic models, it is apparent that these parameters are not directly related to geographical locations which mean problems for model integration with a GIS. The COM integration method is thus adopted, as explained below and illustrated in Figure 1. In the modeling of data management, not only are high resolution remote sensing images and vector graphic data effectively stored but also the input parameters of the epidemic models stored as attributes in the spatial database. The spatial data management technology of GIS can offer a better data interface for model computing in this case. The mathematical model can be integrated into a GIS with COM and the model computing results stored in the same epidemic spatial database with reference to geographic locations. The simulated results may also be visualized in geographic distributional patterns.

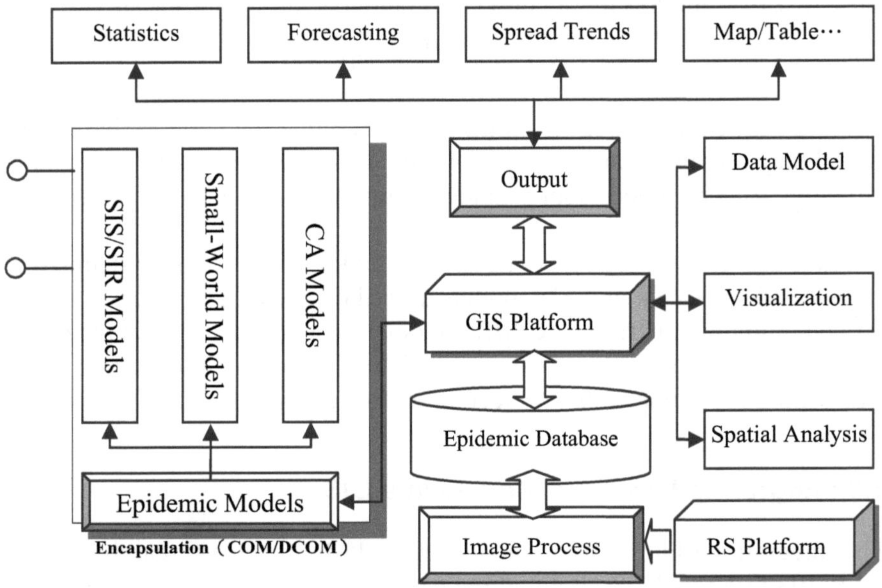

Fig. 1. The integration framework of COM GIS with epidemic spread models

3 · System Development and Implementation

3.1 System Development

The epidemic prevention SDSS involves the integration of GPS, COM GIS, Remote Sensing, Network Communication, Spatial and Temporal Modeling technologies, to support data of multiple scales and formats. The proposed system should also possess the following functions: spatial data editing and processing, spatial modeling, epidemic information querying and publishing, epidemic dynamic surveillance and management, medical resource allocation and facilities planning, epidemic data maintenance, system protection, and user right management functions[8]. The software development setting included the following components:

- Techniques: COM, ASP, Web Service, XML/SOAP;
- Modes: Mixed C/S and B/S;
- Tools: Microsoft Visual Studio6.0;
- Development Platforms

  - Client Side: ESRI ArcObjects, MapObjects, ArcGIS Server and ArcGIS Engine;
  - Server Side: ESRI ArcGIS Server and ArcIMS;
- Application Platforms
  - Client Side: ArcGIS Desktop and IE Viewer as well as ArcObjects, MapObjects and ArcGIS Engine Client;
  - Server Side: ESRI ArcIMS and ArcGIS Server;
- Database: Oracle9i, SQL Server + ArcSDE

3.2 System Structure

The epidemic prevention SDSS comprises several subsystems: (i) data acquisition (including GPS, GIS and Remote Sensing technologies, etc.), (ii) network communications, (iii) model base, (iv) epidemic spatial database, (v) epidemic information query and statistical analysis, (vi) real time epidemic surveillance, (vii) epidemic spatial modeling and decision support, and (viii) epidemic spatial information publishing. The systems structure is as shown in Figure 2[12].

The method of accessing epidemic spatial database varies according to the kinds of epidemic spatial information. ArcSDE, a spatial data accessing middleware from ESRI, is used to retrieve spatial and epidemic data for query, display and record keeping. However, Oracle Objects of the OLE (OO4O) technology from the Oracle Corporation is more suited for retrieving non-spatial data. The OO4O is a COM technology based API for database connection and has a special optimized design for the Oracle Database Systems. Through calling the COM objects and instances of class library offered by OO4O, the client application programs may connect seamlessly to the Oracle database, and the connection speed is faster than that of ADO and other database connection technologies. In addition, the database operational functions offered by the OO4O technology are also more powerful and plentiful.

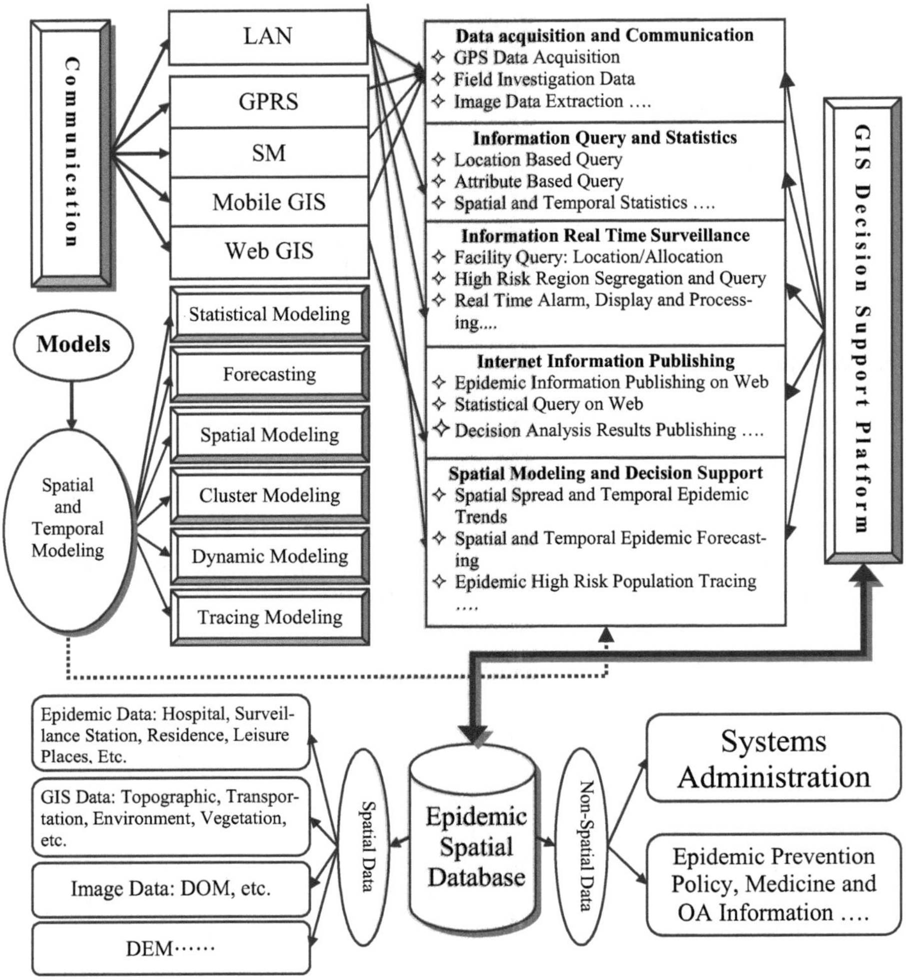

Fig. 2. General structure of the Epidemic SDSS

3.3 Epidemic Spatial Database Construction

With regard to data requirements of the epidemic models and functional characteristics of the decision platform, the following data types are essential: (i) graphic data (including multi-Scale Vector Map Data, GPS located hot spot data and areas of interest extracted from high resolution images, etc.), (ii) epidemic attribute data, and (iii) relevant social and economic data. These data may be easily stored in an epidemic spatial database (such as the Oracle or SQL Sever) through the use of a Spatial Database Engine (as in ESRI ArcSDE)[10]. Data requirements for the epidemic prevention decision support spatial information systems are as listed in Table 1.

Table 1. A table of data requirements for the epidemic decision support spatial information systems

Data Types		Data Names	Scales	Data Usages
Spatial Data	Geographic Data	Basic Maps	1 : 10000 or above	As Display Background
		Buildings	1 : 500 to 1 : 10000	Include Related Attributes
		Roads	1 : 500 to 1 : 10000	Include Road Signs and Video Cam
	Epidemic Thematic Data	Epidemic Cases	1 : 10000	Epidemic Distribution and Attributes
		Facilities	1 : 500 to 1 : 10000	Include Administrative Jurisdictions, Hospitals, CDC, Etc.
		Map Signs	Scale Free	Include Geographic and Epidemic Thematic Map Signs
		Related Information	1 : 500 to 1 : 10000	Key Epidemic Sites : Net Bar, Schools, Transportation Stations, Airports, Water Sources and Monitoring Sites, Etc.
Non-Spatial Data	Systems Management Data	User Management	User Name, Password and User Rights, Etc.	
		Log Data	Systems Access Time and Maintenance Status	
	Systems Dictionary	Matching Table	Epidemic Case Type Matching Tables	
		Translation Table	Translation Table for Items Names, Map Layer Names, and Item Values, Etc.	
	Related Information	Related Table	Used for more related attribute information management.	
		MIS/OA	Epidemic Attribute Data	

3.4 Technical Implementation

The technical implementation of the system mainly concerns COM GIS and relevant software development technologies including ArcObjects, ArcIMS, Visual C++6.0, Visual Basic6.0, Visual InterDev6.0, ASP, Oracle9i + ArcSDE8.X, Decision Analysis Models, et cetera. The COM GIS technology plays a key role in integrating epidemic models with GIS functions and making analytical results spatial and visual. The system components and key technologies are illustrated in Figure 3.

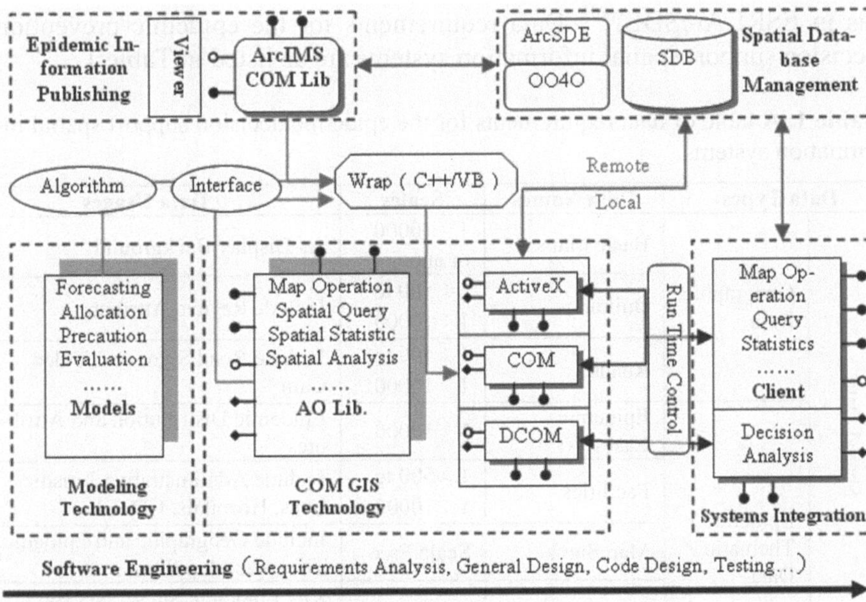

Fig. 3. System components and key technologies

It is recognized readily from Figure 3 that the professional function development based on AO is the key technology for system implementation. To illustrate the basic principles and methods for the system development, the following discussions offer sample codes for a component function of the epidemic case precaution using the VC and VB development tools.

1. Assign a name for the component function and its interface as shown in Figure 4, which includes one attribute and two methods.

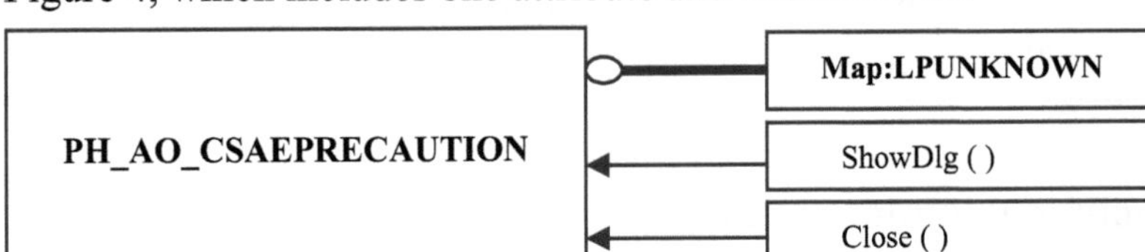

Fig. 4. An interface map of the component function on epidemic case precaution

Implementation in the VC environment.

- Create an Interface Class: Add a new ATL project named PH_AO_CasePrecaution to realize the above attribute and methods through adding ATL Simple Objects.
- Introduce AO component library: Introduce AO core components in StdAfx.h with the following codes:

```
#im-
port"D:\arcgis\arcexe83\Bin\esriCore.olb"raw_interfaces_only,raw_native_types,no_namespace,na
med_guidsexclude("OLE_HANDLE","OLE_COLOR")
```

- Introduce OO4O class library (Oip9.tlb): Put oip9.h into the present project class.
- Call the introduced objects: according to the project, using C++ intelligent pointersto call AO component interfaces OO4O access objects. The following contains codes for executing a spatial query:

```
        IFeatureCursorPtr       _pFeatureCur;       //spatial database pointers;
        IFeaturePtr             _pFeature;          //map layer features;
        IGeometryPtr            _pGeometry;         //feature geometric type;
//create spatial query conditions
        ISpatialFilterPtr _pSpatialFilter(CLSID_SpatialFilter);
// SQL query process
While(if record is not vacant)
{
                if(FAILED(_pSpatialFilter->put_WhereClause("AJLX='AIDS'")))return false;
                _pFeatureCur->NextFeature(&_pFeature);
                if(_pFeature==NULL) break;
                _pFeature->get_Shape(&_pGeometry);
        _pSpatialFilter->putref_Geometry(_pGeometry);
        _pSpatialFilter->put_GeometryField(GeometryFld);
        _pSpatialFilter->put_SpatialRel(esriSpatialRelContains);
        _pSpatialFilter-
>put_SpatialRelDescription(::_com_util::ConvertStringToBSTR(_T("T********")));
//if the relation is contain
        _pSpatialFilter->put_SpatialRel(esriSpatialRelRelation);
//compute the number of epidemic cases, and make precautions.
……
}
        through defining the following OO4O objects，may access Oracle attributes.
        OraSessionPtr _OraSes;              // Oracle object intelligent pointer
        OraDatabasePtr _OraDB;              // Oracle object intelligent pointer
        OraDynasetPtr _oraDyn;              // Oracle object intelligent pointer
        OraFieldsPtr _OraFds;               // Oracle items object intelligent pointer
//the process for connecting and operating Oracle
// COM initialization
        HRESULT hr = ::CoInitialize(NULL);
        if(FAILED(hr)) ::CoUninitialize();
        //create objects for database working space
        hr = _OraSes.CreateInstance("OracleInProcServer.XOraSession");
        if(FAILED(hr))       return FALSE;
//open database
                this->_OraDB=this->_OraSes->GetOpenDatabase( Database Service Name,
                Username/Password, ORADB_DEFAULT );
//start database transaction process
        this->_OraSes->BeginTrans();
// executing SQL query，store the results into data sets
this->_oraDyn= this->_OraDB->GetCreateDynaset("select * from layername ORDER BY
LAYERINDEX",ORADB_DEFAULT,&vtMissing);
//get present items, start to operate Oracle table
        _OraFds=this->_OraDset->GetFields();
……
```

2. Integrating GIS components by using VB codes.

```
'create PH_AO_CASEPRECAUTION objects
Dim m_Caseprecaution As PH_AO_CASEPRECAUTIONLib.AO_CasePrecaution
'calling PH_AO_CASEPRECAUTION attributes and methods，realizing the functions
Set m_ CasePrecaution = New PH_AO_CASEPRECAUTIONLib.AO_CasePrecaution
m_ CasePrecaution.Map = MapControl1.Map
m_ CasePrecaution.ShowDlg
```

3.5 Integration of Systems Function

Besides the basic capabilities of epidemic data query and statistical analysis, the epidemic SDSS offers a list of professional and advanced functions (Table 2). These include a query function for locating medical facilities, location and allocation modeling of medical resources, dynamic zoning of high risk and quarantine areas, the publishing of real time surveillance information, as well as epidemic prevention spatial decision analysis functions. The structural organization of functions for the epidemic SDSS is outlined in Table 2.

The epidemic SDSS is tested using the AIDS monitoring data in the Yunnan Province of China as a case study. Figure 5 displays results of the statistical analyses of the data as well as spatial distributional patterns of AIDS cases in Yunnan [14]. The systems interfaces are also illustrated in the figure.

The epidemic SDSS offers various analytical and visualization functions including thematic maps, statistical graphs/charts and tables. These functions enable displays of the spatial distributional patterns of an epidemic disease and, in conjunction with the integrated epidemic spread models, may reveal the spreading tendency of the disease. The epidemic SDSS provides information beneficial to relevant government and public health institutions that facilitates them to make informed decisions on control and intervention measures.

Table 2. A chart of systems functions

Sequence	Function Classification	Function Descriptions
1	Basic GIS Functions	Include User Management, Data Management, Map Operation, Map Editing, Map Query, etc.
2	Epidemic Information Query	Epidemic Case Query
		Other Relevant Thematic Information Query
3	Epidemic Information Statistics	Epidemic Case Statistics , Spatial and Temporal Distribution Analysis
		Other Relevant Information Statistics
4	Epidemic Information Analysis	Network Analysis
		Precaution Analysis according to the Epidemic Case Numbers
		Medical Resource Location/Allocation
		What-if Scenarios
		Epidemic Case Forecasting According to the Forecasting Models
		3D Police Information Analysis
		Trend Surface Analysis
		Disaster Evaluation and Recovering Planning
5	Decision Analysis Results Output	Analysis Results Publishing Based on Web GIS
		JPG, BMP Image Output
		Print Output

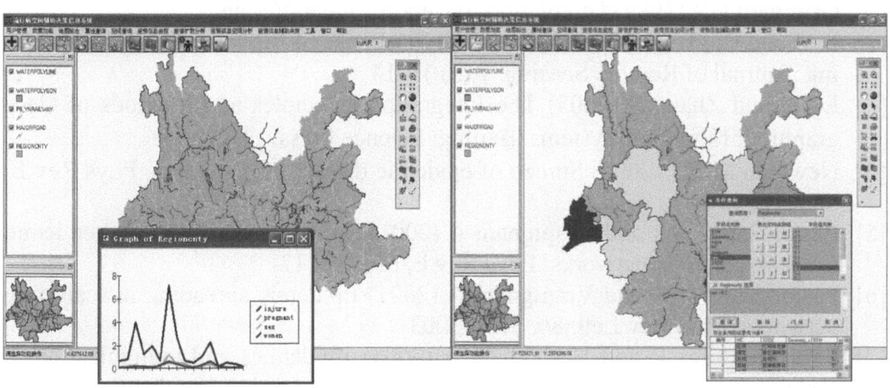

Fig. 4. The system interface and results of the Epidemic SDSS illustrating AIDS distribution in the Yunnan Province of China

4 Conclusion

Modern technological developments have strengthened man's ability to prevent epidemic diseases to a large extent. However, the human race is still facing problems in understanding disease spread at various spatial scales (i.e. local, regional, and global), recognizing highly contagious diseases, and effectively detecting outbreaks before they happen. This chapter has demonstrated the values of spatial information technologies in an epidemic SDSS through the integration of epidemic spread models in a GIS. The epidemic SDSS can assist the government and relevant public health institutions manage disease data spatially and temporally such that disease data can be correlated with environmental and social economic factors. The overall aim is to tap benefits of spatial information technologies in delivering SDSS to effectively monitor and prevent disease spread.

Acknowledgement

This research is financially supported by the National Science Foundation of China under Grant No. 30560135 and the Yunnan Provincial Science Foundation under Grant No.2004D0032M.

Reference

[1] Grimmett GR (1989) Percolation. Berlin: Springer-Verlag
[2] Li Z and Cui H (2003) SI Models and Piecewise SI Model on SARS Forecasting. Journal of Remote Sensing. 7 (5):10-14
[3] Lu G and Zhang S (2003) The Integration Principles and Methods of Geographic Information Systems. Beijing: Science Press. 1~345 pp
[4] Newman M E J (2002) Spread of epidemic disease on networks. Phys Rev E, 66, 016128
[5] Pastor-Satorras R and Vespignani A (2001) Epidemic dynamics and endemic states in complex networks. Phys Rev E, 63, 066117
[6] Pastor-Satorras R and Vespignani A (2001) Epidemic spreading in scale-free networks. Phys Rev Lett, 86, 3200-3203
[7] Pastor-Satorras R and Vespignani A (2003) Epidemics and immunization in scale-free networks. In Bornholdt S, Schuster HG (eds.) Handbook of graph and networks. Berlin: Wiley-VCH
[8] Qi X and Lu F (2004) GIS Application and Development in Epidemiology. Chinese Journal of Epidemiology. 25(11):83-85

[9] Sander LM and Warren CP (2002) Percolation on disordered networks as a model for epidemics. Math Biosci, 180:293-305
[10] Wen L and Xu D (2004) The Construction of Malaria Precaution Systems Based on GIS. Chinese Journal of Public Health 20(5):118-119
[11] Xu Z and Z. (2004) GIS Applications in Epidemiology. Chinese Journal of Public Health. 20(8):130-132
[12] Zhong K. and H. (2004) The Design and Implementation of Epidemic Disease Emergency Response Decision Support Systems Based on GIS/GSM. Journal of Surveying and Mapping Science. 2004.1:36-39
[13] Zhou Y. and T. (2003) The Mathematical Models for SARS Spread Forecasting. Chinese Journal of Engineering Mathematics. 20(S2):33-38
[14] Lu L and Jia M. (2005) 1989-2004 AIDS Epidemic Analysis of Yunnan Province, Chinese Journal of AIDS & STD, 2005.9:178-182

Development of a Cross-Domain Web-based GIS Platform to Support Surveillance and Control of Communicable Diseases

Cheong-wai Tsoi

Hong Kong SAR, China

Abstract: Ever since the outbreak of SARS and the recent re-emergence of Avian Flu around the world, there has been a compelling urgency in establishing an adequate mechanism to support the monitoring and control of infectious diseases. Much has been discussed on the application of GIS in the field of public health service. However, the fundamental problem of dispersed data and the reality that health information systems and GIS are mostly in isolation have not been sufficiently addressed. In such a context, the effective use of GIS by public health organizations or government agencies in fostering the goal of safeguarding the general public against the spread of fatal diseases presents itself as a major challenge. This chapter attempts to review two successful models that have adopted a web-based GIS technology to overcome these real world constraints. One of them has improved the data dissemination efficiency in the combat against a particular vector borne disease, while the other aims at developing a cross-domain solution which gives rise to a near real-time tracking of over 30 types of infectious diseases. The examination of the design and benefits of such a web-based GIS platform concludes that through transcending across institutional boundaries in a collaborative way, the capability of monitoring and control of infectious diseases can be greatly enhanced.

Keywords: GIH, GIS, geospatial, dengue fever, infectious diseases, surveillance, health.

1. INTRODUCTION

While geospatial information technology has been applied widely in many disciplines, its full potential has not yet been entirely unleashed to bolster public health management. One of the fundamental issues affecting the application of geospatial information technology in this area is the amount and availability of data. Data that are useful for disease investigation and control are often derived from different sources: case data from patients, private medical practitioners, clinics or hospitals; environmental data from laboratories or equipment installed in the field; and geographic data from an even much diverse origins like land administration authorities, engineering departments and local district offices. Without these vital data readily accessible, researchers or medical professionals are coerced into performing retrospective analysis of historical data only - an undesirable scenario that is very much the norm today. The multitude of data in itself already presents certain level of difficulties in setting up a Geographic Information System (GIS) for public health management, let alone the creation of appropriate public health surveillance GIS which demands the integration and constant updating of essential spatial and non-spatial data.

Another aspect hampering the utilization of geospatial information technology in the public health sector is that most of the advanced GIS functions in the last decade were provided and tailored solely to standalone GIS software packages. The costs and the amount of special training needed to operate these sophisticated systems have deterred many public health professionals from embracing their use.

2. BACKGROUND

Similar to other governments in the world, many departments in the Hong Kong government have commissioned project-based GIS focusing on their own specific internal needs. As a result, data are scattered around different locations, stored in various formats, have numerous data structures and bear inconsistent definitions. Contrary to the notion of turnkey systems promoted by software vendors, many of these installed departmental level GIS require intricate operating procedures and have therefore remained accessible only to the knowledgeable few who are able to grasp the underlying GIS theories, processing procedures and data quality issues. In addition, reservations about data sharing and clashing work priorities among departmental officers aggravate this problem of data acquisition. Taking all of these into consideration, it has become most strenuous and time consuming, if at all possible, to obtain and consolidate essential data for serving the much needed operations for public health.

To redress the above issues, it is crucial to design and establish a GIS model that is readily accessible and highly user-friendly. Such a model should consist of a common infrastructure providing integrated geospatial data and basic GIS functionalities for users across several organizations.

Further to some coordination and development efforts, the concept of sharing and integration has finally come to fruition in the form of a common GIS platform called the "HKSAR Geospatial Information Hub (GIH)". The GIH is a ground-breaking GIS platform in Hong Kong which adopts the web technologies. It is a one-stop geospatial information portal aiming at improving geographic information service delivery among the government departments. The GIH, initiated and implemented by the Lands Department (LandsD), integrates vast amounts of geospatial data gathered from diverse sources covering the entire territory. It does not only include fundamental mapping data, but also encompasses ortho-rectified aerial photos as well as data relating to land administration, buildings, population, education, transportation, community and other demographics. All data in different formats, structures and geo-referencing systems are collated and integrated into a single geospatial database under a common coordinate system for easy referencing. This chapter uses two examples related to public health issues to introduce this cross-domain web-based GIS platform which was put into practical use since 2004 to address many scenarios in Hong Kong.

3. WEB-BASED GIS TO SUPPORT DENGUE VECTOR SURVEILLANCE

Dengue fever is a re-emerging infectious disease affecting millions of people worldwide and it can be transmitted via the mosquito of species Aedes albopictus. To mitigate the threat of dengue fever, the Food and Environmental Hygiene Department (FEHD) has installed ovitraps to detect monthly larval breeding rates of mosquitoes in strategic locations across 38 regional zones and 30 ports to closely monitor the vector situation. [3] An Inter-departmental Working Group on Pest Prevention and Control has also been set up to oversee the sharing and dissemination of surveillance results to more than 20 bureaus and departments, as well as the implementation of mosquito control measures in premises under the jurisdictions of these bureaus and departments. Information dissemination has not been entirely effective prior to the launching of GIH because results of the ovitrap surveillance must be manually marked on scanned maps before dispatching to individual departments through faxes or emails; a tedious and inefficient process for both senders and recipients. Worse still, the amount of efforts put into the preparation of the bulky reference map as image files has little bearing on the data quality. The FEHD has therefore forged a close partnership with the LandsD to explore possibilities of improving the operation of dengue vector surveillance. The Web-based GIS technology was identified as the means to enhance support among all parties concerned with an aim to better protecting the public.

The FEHD took advantage of the GIH platform as outlined above in a bid to upgrade efficiency in its surveillance processes. Accurate mapping of Dengue disease vector information on top of rich geospatial information readily available from the GIH becomes a powerful asset for environmental health management. To meet with the objective of improving competence in dengue vector control, data exchange standard has been established between the FEHD and the LandsD. New functions have since been further developed on the GIH for the display of ovitrap positions as well as facilitating query on ovitrap counts (i.e. the number of eggs or larvae in each ovitrap). Furthermore, the enhanced GIH provides functions to display monthly ovitrap index, allow the visualization of spatial distribution, and to analyze the severity of the vector problem over time (Fig. 1). It also helps to identify trends and patterns and to alert where there is a need to target extra resources in environmental cleaning.

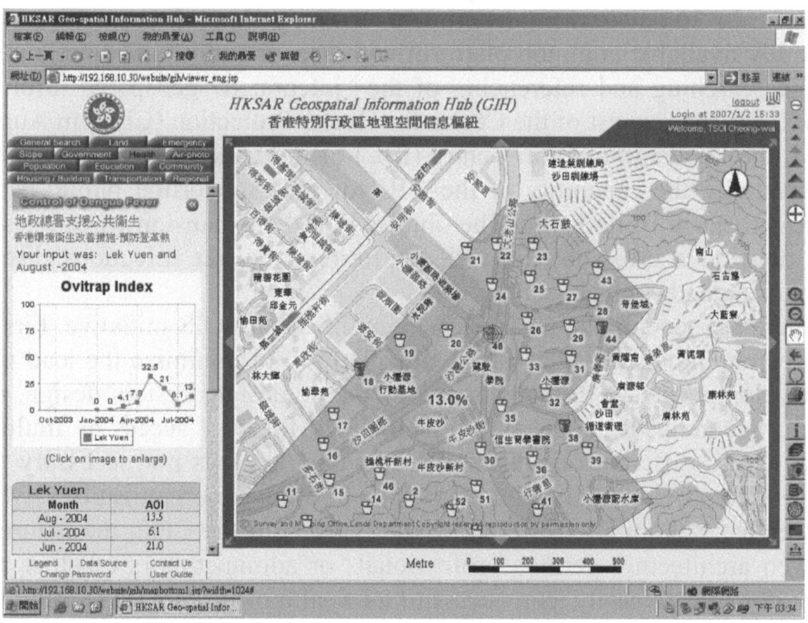

Fig. 1 GIH showing ovitrap locations and indices; ovitraps with a positive count of Aedes Albopictus are symbolized in a different color. (*Simulated data are used*)

4. WEB-BASED GIS TO SUPPORT MONITORING OF INFECTIOUS DISEASES

The government and public at large have become acutely aware of the importance of timely and accurate information following the outbreak of SARS. Such information needs to be reliable and readily available for decision making and swift responses. With the re-emergence of the Avian Flu around the world in late 2005, there is an urgent need for the Government of HKSAR to put into place mechanisms to provide decision makers a quick overview of the changing situation while allowing medical professionals simultaneous access to essential data for monitoring and analyzing the changing conditions.

Building a sophisticated public health GIS from scratch takes time and significant human and financial resources. Besides project management and procurement, much effort is required on system analysis and design, data collection, data conversion, geo-coding, data integration, application development, system tuning and testing, as well as seeking the involvement of all parties concerned. By 2005, the GIH has been put into practical use by over 30 government departments in Hong Kong. It has evolved into a readily available common platform with user-friendly GIS functions for online searching and referencing of the "integrated" geospatial information. Upon the request of the Centre for Health Protection (CHP) in August 2005, the Land Information Centre (LIC) of the LandsD began to expand substantially the capability of the GIH towards supporting public health management in Hong Kong. The objective of the expansion was to design and develop a cross-domain GIS platform to bolster infectious disease monitoring and control.

There are several characteristics about this health GIS initiative. Firstly, it rides on an existing web-based GIS platform to minimize the lead time in development, data conversion, cartographic design, system design, and so on. Secondly, it adopts a web technology for easy access by multiple users across different organizational domains (i.e. it is not necessary that the service provider and end users belong to the same organization). While the service was initially offered by the LandsD to the CHP, the nature of its web architecture permits professionals or administrators of other network domains to easily gain access to its wealth of resources. Thirdly, this cross-domain GIS project possesses abilities to support outbreak investigation by medical professionals. This quality is achieved through a mechanism that enables cross-domain data updates at frequent intervals. Disease case data are first validated based on a schema of data type and allowable values. They are then automatically converted and uploaded into a centralized geospatial database in the GIH back-end servers located in the

LandsD. Web mapping modules in the GIH then combines the disease related data with the geospatial data to enable, within a short period of time, spatial and non-spatial queries and to allow the display of infectious disease incidence over any geographic coverage in Hong Kong (Fig. 2).

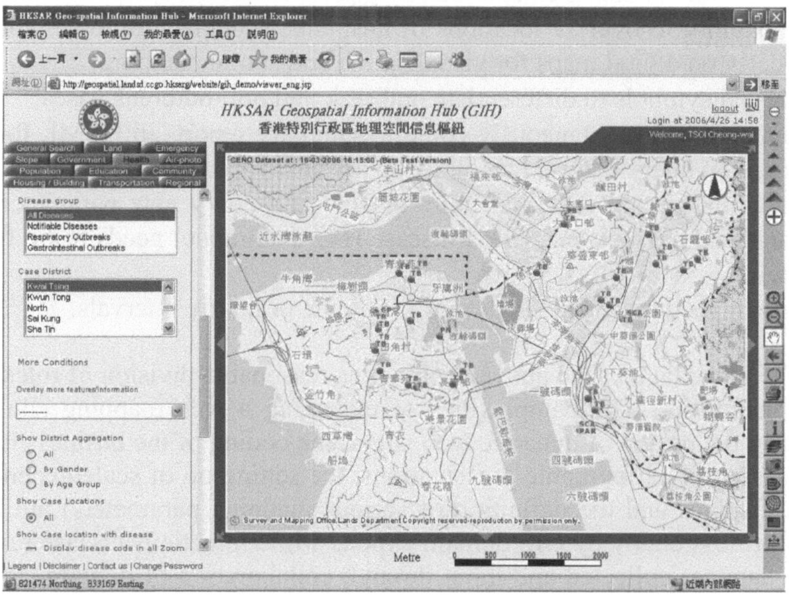

Fig. 2 GIH showing the distribution of an infectious disease based on specific query criteria (*Simulated data are used*)

A list of 31 types of infectious diseases (e.g. mumps, severe acute respiratory syndrome (SARS), influenza A (H5), chickenpox, measles, meningococcal infections, etc) is specified in the First Schedule of the Quarantine and Prevention of Disease Ordinance (Cap. 141) in Hong Kong. Registered medical practitioners, as required by law, must give notification of suspected or confirmed cases of diseases on the list [2]. Some key functions being engineered for this public health GIS platform to monitor these notifiable diseases are as follows:

- Capability to display spatial distribution of all notifiable diseases reported by registered medical practitioners and hospitals;
- Function to permit query on disease case data based on date of notification, type, case district, etc;
- Ability to display cases based on geo-coded addresses and their corresponding disease codes on maps;
- Means to display disease cases based on specified disease groupings;
- Function to provide tabular summaries based on query criteria;

- Capability to view case attributes after performing spatial query (e.g. case district, home address, office name, office address, age, sex, hospital attended, etc.);
- Function to display aggregated number of disease cases in different districts with bar charts categorized by gender or age group;
- Capability to overlay locations of hospitals, schools and homes for the elderly on digital maps for visualization and action planning;
- Special symbols to differentiate outbreak and non-outbreak cases;
- Function to instantaneously compile summary reports in HTML format based on query criteria for follow up actions;
- Flexibility to process, integrate, display and query new parameters or data fields associated with a disease (i.e. data model needs to be flexible);
- Infectious disease data update at 15 minute or hourly intervals.

The advantage of the above approach is reasonable division of roles and responsibilities. The complexities of managing a web mapping site (including the design of dynamic web pages, the coding of the business logic, the design of cartographic presentations, the setting up of scale dependent map displays and the definition of various mapping parameters) are handled professionally by GIS domain experts while infectious disease related case data are collected and input through collaboration of domain experts in the medical profession.

5. DESIGN MODEL AND DEVELOPMENT TECHNOLOGIES

In addition to server-side GIS and relational database software, a variety of computer development languages and technologies (e.g. Java, JSP, JavaScript, DHTML, XML, COM, etc.) were utilized to support the development of this cross-domain web-based GIS platform. A scalable solution was chosen to cater for the possible increase in capacity needs. The GIH adopts a multi-tiered architecture in its overall system design [6]. In the data tier, geospatial data are updated and organized in an enterprise geospatial database optimized for retrieval. The database organizes geospatial data in predefined feature classes and database tables for efficient query, retrieval and display. Conversely, spatial queries are processed through a spatial data engine on a relational database. In the business logic tier, the GIH consists of web servers, servlet connectors, map application servers and map spatial servers. In the presentation tier, a thin client approach is adopted. To enable the contents reaching as many people as possible across the government network, the presentation tier takes on a minimalist approach where only a simple HTML viewer or an Internet browser supporting DHTML is required.

Data update and data retrieval are two major issues in designing a GIS. There are different approaches to system-to-system integration. While Web Services approach has some merits in certain situations, it was not specifically adopted to support objectives of this health GIS implementation. The key concerns have been speed and performance, ease of use for a diverse group of users, service availability and additional enhancement needed on existing computer systems in each organization. In the current implementation approach, pest control and medical professionals can readily access, view, and study geographic as well as public health related data. This widespread access is achieved through extensive methods developed on the expanded GIH for data query and map interaction.

6. BENEFITS AND RESULTS

In the battle against dengue fever, the adoption of a cross-domain web-based GIS allows case officers of concerned departments to access at any time up-to-date surveillance information on dengue vector through the government intranet (Fig. 3). Officers are able to retrieve both current and previous readings of ovitraps in the form of high quality digital maps showing the locations of ovitraps with positive counts. Compared with the conventional means, the current form of information access is far more comprehensive and integrated. The web-based GIS details the types of buildings, exact locations of positive ovitraps and the population affected. Trend analysis where line graphs against time are generated instantaneously from ovitrap index data can be performed with simple mouse clicks. This ease of data access facilitates easy visual interpretation of ovitrap index trend in a particular district by the relevant parties.

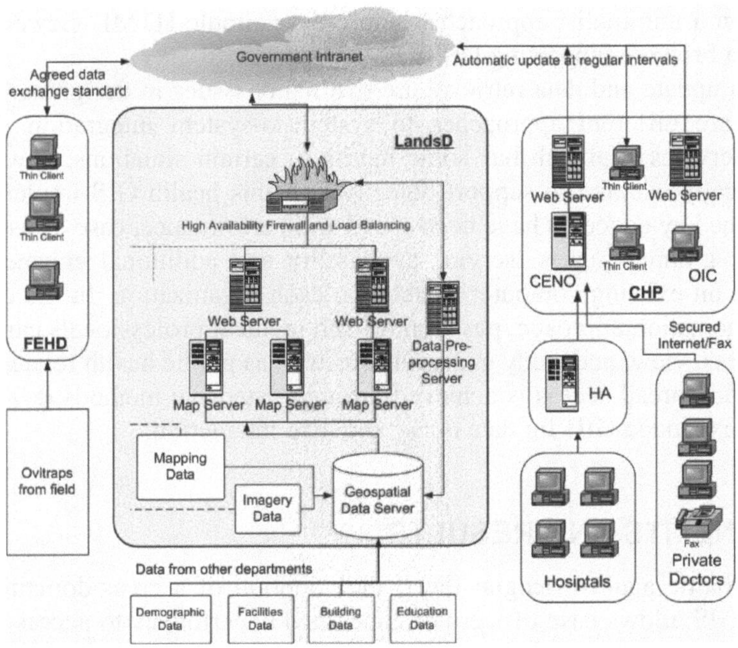

Fig. 3 A schematic diagram illustrating the relationship of different domains

The new approach on the GIH also provides a real-time spatial overlay capability. Given the average flight range of the Aedes mosquitoes is approximately 100 meters; this new GIS solution enables spatial analysis on demand and in real time to enlist buildings within 100m radius of ovitraps with positive readings (Fig. 4). Going one step further, the GIH categorizes these buildings into homes for the elderly, schools and hospitals. This breakdown helps relevant offices to visualize better the impact of Aedes mosquitoes to the surrounding schools, homes for the elderly and hospitals, thus enabling quick follow up actions. The joint effort of both the LandsD and the FEHD greatly facilitates the implementation of mosquito control measures in a timely and highly coordinated manner. For example, the use of the GIH platform by the Home Affairs Department means that the surveillance results can be passed in turn to partners in the community, such as District Councils and Area Committees. Property management of housing estates, schools, construction sites and so on can be informed as well to take prompt and specific measures to contain mosquito problems in their own premises. Since the operation of GIH, more than 20 bureaus and departments are better equipped to carry out dengue vector control. This new model has proven successful– in terms of saving staff time, enhancing efficiency and facilitating coordinated responses in Hong Kong.

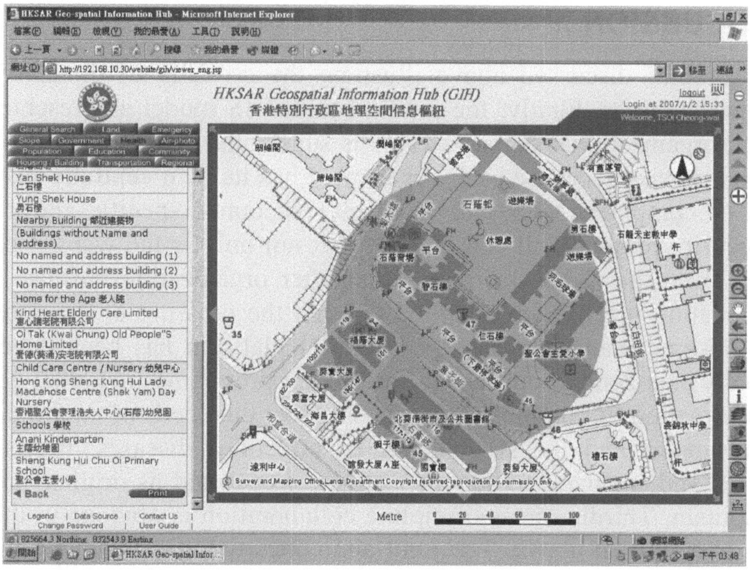

Fig. 4 Spatial analysis function in GIH. All buildings within 100 m of a selected positive ovitrap are highlighted on the map window using a different color. A list of buildings with higher possibilities of being affected by Aedes mosquitoes are shown on the left. These buildings are grouped by categories to facilitate response action. (*Simulated data are used*)

The second public health GIS initiative has a much wider scope and complexity than the one described above. Spatial analysis of data on infectious diseases cannot be performed rapidly or directly and with ease by medical professionals in the past because the task requires appropriate equipment, coupled with essential geographic data and GIS expertise. The development of new functions related to surveillance of infectious diseases in the GIH and the incorporation of disease case data have enabled statutory notifiable diseases be investigated and analyzed from the spatial perspective in a near real-time manner by medical professionals simply through an Internet browser. While simple correlation analyses may have been performed previously, the new approach provides a convenient means to conduct temporal and spatial queries. Spatial distribution and disease clusters can be identified more readily. This change has made possible the provision of crucial information to support early and better coordinated response actions in case of possible outbreaks of infectious diseases in the community. The GIH, with both geographic and demographic data, provides an important tool for effective public health monitoring.

7. DISCUSSION

The examples discussed above illustrate the crossing of domains from three perspectives. Firstly, the public health GIS model as presented has crossed organizational domains. This is substantial achievement from the IT point of view because each department has its own centrally-managed group of computers with individual user accounts, specific applications and security protocols. It can be seen that not only is the web-based GIS service from the LandsD offered to another organization but also is the data updating mechanism operated through the crossing of domains in a near real-time manner. Secondly, the GIH platform has been developed originally as a general purpose infrastructure providing integrated geospatial information of common interests. To meet the needs of environmental health support and the surveillance of infectious diseases, it has been modified and enhanced to adapt specific tasks beyond its original application domain. Thirdly, cross-discipline collaboration is obvious in the scenarios exemplified above. The joint and concerted efforts among a range of domain experts (e.g. pest control, medical and GIS professionals) have permitted substantial progress made within a limited period of time, benefiting the community as a whole.

The cross-domain web-based GIS supporting public health management could not be realized without spatially and temporally accurate and complete data. The SARS outbreak in 2003 resulted in a relatively well established health-related incident reporting mechanism to be put in place in Hong Kong. Supported by the Prevention of Disease Ordinance, vital and basic information of infectious disease incidents were collected, converted and organized into a digital format by the government. Such reporting mechanism has improved the availability and timeliness of incident data as well as geographic data, thus reducing the chance of spatial uncertainty or misleading results [4, 5]. This move has also improved the quality of GIS analysis via the web-based GIS platform for formulating emergency responses.

A comprehensive policy on Spatial Data Infrastructure (SDI) does not exist in Hong Kong at the present time. Despite this inadequacy, the Government of the Hong Kong SAR has managed to adopt a pragmatic bottom-up incremental approach in building this health GIS. More than being a surveillance system established for early detection and mitigation of health problems, the current model can also serve as a platform for illustrating the need of formulation and revision of relevant policies – policies related to "sound and comprehensive spatial health data and information infrastructure." [1]

The concept of GIH is indeed a step beyond establishing a one-stop portal simply for searching and unearthing data. The GIH initiative can be viewed as a business re-engineering process within the government to promote collaboration amongst diverse disciplines. It enables government officers to access a multitude of geospatial information on the fly. When applied to public health, it dramatically improves information flow and relieves case officers from time consuming negotiations and labor intensive data searches from several locations. More importantly, it reduces duplication of efforts in data search and conversion while eliminating the processes of manual handling of paper-based geographic information prior to actual usage. The integration of information from multiple sources into a unified platform allows greater flexibility in exploring new patterns or phenomena. Health professionals may gain better insights about a situation never before encountered or easily tackled.

8. CONCLUSION

Hong Kong is a densely populated city, bordering mainland China, bustling with local activities and has close connections with international communities. In the face of public health threats, swift and appropriate responses are essential to protecting its citizens as well as a pre-requisite for being a responsible member of the global village. The world did not have adequate preparation for the SARS outbreak in 2003. However, we have access nowadays to more information, better technologies, and advancement in scientific research. The main challenge in the application of GIS for health is a practical deployment model that is compatible with realities of the working environment. GIS does not purport to solve all public health problems but it can be a very useful tool in the discovery of patterns, trends or unusual phenomena over a geographic area. It can also facilitate risk assessment of vector-borne diseases and the continuous monitoring and management of infectious diseases. The examples presented in this chapter not only aim at consolidating data spanning across three organizational domains but also demonstrate thriving collaboration among three seemingly different professions. Despite the many factors affecting decision making on public health issues, the web-based GIS can better facilitate the surveillance and control of infectious diseases, thus giving rise to more effective and coordinated responses. The model described above illustrates the successful implementation of GIS in respect of public health issues by resolving the problems of consolidating diverse data dispersed across divided domains.

Reference:

1. Boulos, MNK. (2004) Towards evidence-based, GIS-driven national spatial health information infrastructure and surveillance services in the United Kingdom. International Journal of Health Geographics 2004, 3:1
2. Centre for Health Protection. Notification of Infectious Diseases http://www.chp.gov.hk/notification.asp?lang=en&pid=13&id=33 Accessed on April 7, 2006
3. Food and Environmental Hygiene Department. Ovitrap Survey in Community and Port Areas http://www.fehd.gov.hk/safefood/pestnewsletter200602.html Accessed on April 1, 2006
4. Jacquez, GM. (1998) GIS as an Enabling Technology. In GIS and Health (Eds. Gatrell A, Loytonen M.) Philadelphia: Taylor and Francis, Inc; 1998: pp.17-28
5. Oppong, JR. (1999) Data Problem in GIS and Health. Proceedings of Health and Environment Workshop 4: Health Research Methods and Data. 22 July to 25 July 1999, Turku, Finland. http://geog.queensu.ca/h_and_e/healthandenvir/Finland Workshop Papers/OPPONG.DOC Accessed on March 29, 2006
6. Tsoi, Cheong-wai (2004) HKSAR Geospatial Information Hub – Design, Development and Implementation. Proceedings of The 4th Across-the-Straits Geomatics Conference. 16 August to 18 August 2004, Changchun, Jilin Province, China [CD-ROM]

Acknowledgements

The author thanks his colleagues in the Web GIS Development Unit of the Land Information Centre who have contributed to the development and implementation of the GIH service.

Remark

The author works in the Land Information Centre of the Lands Department in Hong Kong SAR, China. The views expressed in this chapter represent the views of the author and do not necessarily reflect the views of the Lands Department or the Hong Kong SAR Government.

A GIS Application for Modeling Accessibility to Health Care Centers in Jeddah, Saudi Arabia

Abdulkader Murad

King Abdulaziz University, Jeddah, Saudi Arabia

Abstract Defining health care accessibility is one of the major tasks that should be covered by health planners. This task can be implemented based on spatial and non-spatial factors. The former takes into account distances to the nearest health centers and provider-to-population ratios. The latter non-spatial factors cover issues such as social class of patients, income and age of health demand. Both factors depend on the use and analysis of spatial data such as the location of health centers, health care catchment areas, and population zones. All of these data can be handled and integrated using a GIS. This chapter presents a GIS based application for defining spatial accessibility to health centers in Jeddah city, Saudi Arabia. This application produces three accessibility models: (i) distance to the nearest health centre, (ii) health centre-to-population ratio, and (iii) a combined health care accessibility indicator based on spatial interaction techniques. Each one of these models aggregates the city into accessibility zones. The resulted zones can be used by health planners in deciding whether current health facilities cover the entire city on not. The ArcGIS Spatial Analyst and Network Analyst modules are used in this application. The former is used to produce a raster-based accessibility model for health centers while the latter is used to model health care accessibility using spatial interaction techniques to predict the flow of patients from their residential areas to the nearest health centers.

Keywords: accessibility, health care, GIS, Spatial Analyst, Network Analyst, spatial interaction.

1. Introduction

Defining accessibility to health centers is one important task for health planners. This task requires large amounts of data collected from different sources including the location of health centers, district-based population, and distance between health centers and demand locations. These different data sets can be handled and manipulated spatially using a Geographical Information System (GIS) whose main feature is the ability to analyze spatial data using different functions.

The aim of this chapter is to use this novel technology on health centers located in Jeddah city, Saudi Arabia. The subject of health care planning includes several spatial aspects such as defining health status or identifying patterns of diseases. This study will focus on the aspect of spatial accessibility to health centers and will use GIS to identify levels of accessibility to these centers. The first part of the chapter defines accessibility and its measures. The second part discusses GIS tools and functions used to define accessibility. The third part of this chapter presents a GIS application created for Jeddah city. This application focuses on health centers and defines their spatial accessibility based on distances and demand centers.

2. Background

2.1. Defining Accessibility

The simplest definition of accessibility of a given location is in terms of how easy it is to get there. Moseley (1979) indicated that 'accessible' is meant crudely as the degree to which something is 'get-at-able'. The reasons that some places are 'inaccessible' or 'difficult to get at' may vary as a result of physical and social dimensions of accessibility. The former refers to the ability to command transportation network facilities needed for reaching supply locations at suitable times. The latter focuses on the individuals who must fulfill certain requirements (in terms of age or ability to pay or overcome the barrier) to reach the supply location or destination (ibid).

Accordingly, what is accessible by a private car may be inaccessible by public transports, and what is within an accessible walking distance for the young may be effectively inaccessible for senior citizens (Jons and Van Eck 1990).

2.2. Accessibility Measures

Phillips (1990) suggested that health care accessibility can be divided into two categories called revealed and potential accessibilities. The first category deals with the actual use of health care services while the second category concerns the geographical patterns and aggregate supply of medical care resources (Thouez et al. 1988). Luo (2004) further divided these two categories into spatial and non-spatial accessibilities. Spatial accessibility emphasizes the importance of geographic barriers (distance or time) between consumers and providers whereas non-spatial accessibility focuses on non-geographic barriers or facilitators such as social class, income, ethnicity, age, sex, etc (Wang and Luo 2004). Since this chapter is concerned with GIS applications in health care services, the discussion will be focused on the spatial aspect of health care accessibility and its measuring techniques.

The literature on accessibility measures showed a need for quantitative indicators of accessibility for different kinds of public services including health care. Such indicators would serve as instruments in the comparisons of accessibility in different parts of the region and in the evaluation of alternative plans for new service facilities and transportation links (Al-Sahili, et al. 1992). Examples of accessibility indicators are: provider-to-population ratio, distance to the nearest provider, average distance to a set of providers and gravitational models of provider influence (Guagliardo, et al. 2003). Each one of these indicators can be used to evaluate accessibility of health care centers. For example, Luo (2003) computed accessibility to physician locations west of Chicago based on the physician-to-population ratio. The same principle was applied by Wang and Luo (2004) but with added non-spatial factors of health care accessibility, such as age and social class of Illinois, USA. Murad (2004) presented the use of gravity models in defining the flows of population to health centers in Saudi Arabia. The same technique was also found in the works of Birkin, et al. (1996) defining patient flows to health centers in the U.K. The gravity based models produced a combined accessibility indicator of distance and availability, and could provide the most valid measures of spatial accessibility (Guagliado et al. 2003). This chapter presents a modified approach for spatial accessibility measures based on service-to-user ratio and gravity models.

3. GIS Tools for Accessibility

A GIS has many spatial analysis tools that can be applied on different planning problems. These applications include land use analysis, facilities planning and optimization studies. One of these applications is related to defining accessibility of health care facilities in built up areas. A GIS can be of great help for these studies needing to define the level of accessibility for a certain location. The accessibility tools available in a GIS are divided into two main types. The first type is related to the vector data model and the second type to the raster data model.

3.1. Vector Based Tools

GIS accessibility tools for vector data (represented as points, lines and polygons) include simple and advanced functions. For example, the proximity analysis (also known as the buffer analysis) based on distance buffers measured from selected features, can be used to evaluate the accessibility of any location according to the distance factor. Most GIS software possesses this kind of distance-based analytical function to identify, for example, risk areas by delineating zones or buffers around selected features (Grimshaw 2000). Gravity or spatial interaction models are considered as advanced tools for defining accessibility of vector data. These tools are offered by some GIS software such as ArcGIS, and can be used to test or evaluate likely trade areas defined by iso-probability contours (ibid).

3.2. Raster Based Tools

GIS raster data are made of grid cells or pixels (i.e. a square representing a specific portion of an area). For example, a road or a line can be represented in the raster format as a sequence of connected cells. Like the vector data model, raster based accessibility functions are known as distance mapping functions which are divided into straight-line and cost weighted distance functions (McCoy and Jonhston 2001). The straight-line distance function measures the straight-line distance from each cell to the closest source. The cost weighted distance calculates distances based on the cost to traverse any given cell. Both of these functions are considered useful in finding out information such as distances from certain areas to the nearest hospitals which, in turn, determines the level of accessibility of a hospital.

4. GIS for Examining Health Centre Accessibility

The previous sections discussed different accessibility indicators that can be used with GIS for the purpose of measuring accessibility. Three accessibility models based on the following indicators are built in this study: 1. distance to the nearest health centre, 2. health centre-to-population ratio, and 3. a combined health care accessibility indicator. The kinds of data required for the computation are described below.

4.1. The Database

Several data sets in non-digital format were collected and transformed into digital format for input to the ArcGIS software. The created database includes the following datasets:

A: A road network for Jeddah city was created as a GIS line coverage. It describes the pattern of road networks and it includes data about road length and road type. This coverage is useful for health planners because it enables them to understand the spread of a city network and it can be overlaid with a location map of health centers to examine the relationships between the location of health centers and the road type.

B: A point GIS coverage was made to record locations of local health centers in Jeddah city. This coverage includes several useful attributes such as centre size, service area, registered patients, equipments etc. Health centers can be classified on any of these attributes and the results can be used for comparing functional differences of health centers.

C: The third important data for this study are the boundaries or service extents of health centers. This is a polygon coverage dividing Jeddah city into several zones where each zone shows the spatial extent of a health centre. Several attributes are linked to this coverage including centre name, population size, patient size, and area. These data are also useful for health planners because they not only specify the real catchment area of each health centre but also indicate whether the centers can offer adequate service for population of a certain size.

4.2. Proximity to Health Centers

One way of defining accessibility to health centers is by knowing how far patients live from their nearest centers. In this case, distance to provider is the main tool for measuring a centre's accessibility. Based on the planning standards of the Ministry of Municipal and Rural Affairs, every health centre should cover a catchment area extending 2 km radius wide. Accordingly, patients residing within 2 km distance of a health centre are entitled to receiving services from this centre. A GIS is used by the present study to define which parts of the city fall within the 2 km catchment area and which can be labeled as the served parts of Jeddah city.

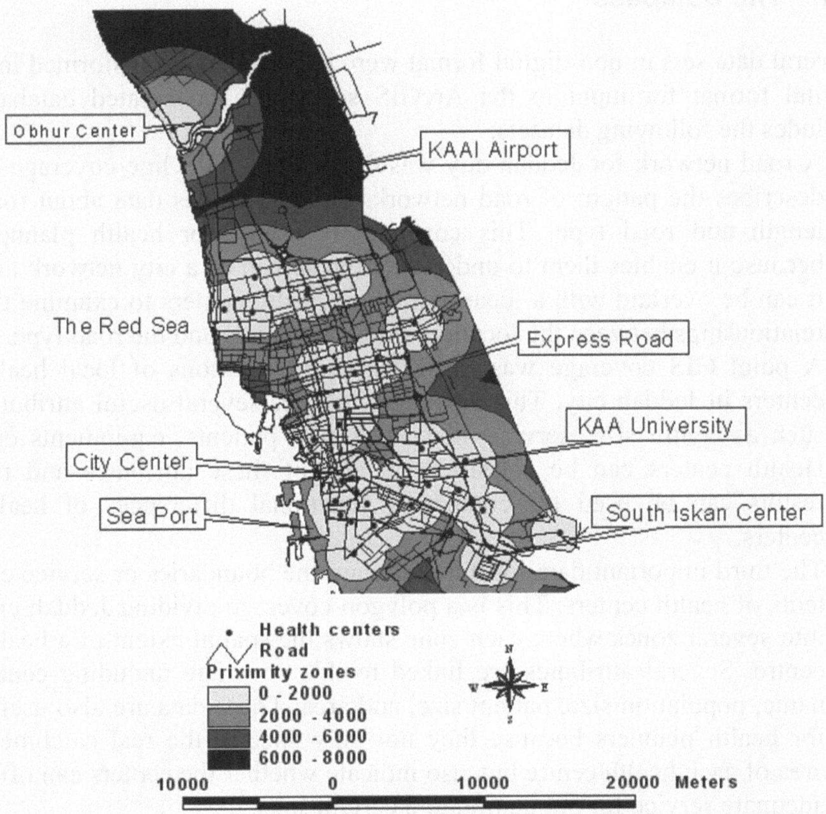

Fig. 1. Proximity to health centers in Jeddah City

May GIS functions can be used to define distances around features in the vector or raster formats. The study selected the ArcGIS Spatial Analyst to calculate distances around the health centers. For this application, the distance function selected to model proximity to the health centers computes an output raster dataset in which the output value at each location is potentially a function of all cells in the input raster data sets (McCoy and Jonhston 2001). The raster output of this function contains the measured distance from every cell to its nearest source. It is measured in projection units or meters in the case of Jeddah city. The distance function are potentially useful for such applications as finding the nearest hospital for an emergency helicopter, or finding the best site for a new school (ibid). However, this study applied the distance function to define distances around every health centre in Jeddah city (Figure 1).

It is clear from Figure 1 that several parts of Jeddah city are located outside the 2 km accessibility zones. These areas are mainly situated north and east of the city with some to the west. It is also clear from this figure that existing health centers must serve a catchment area larger than the standard size. Based on this output, different parts of Jeddah city are assessed as having a low health accessibility service, including AlMohammadia district located north of Jeddah, and Alhamra district west of the city.

Health planners can use this model in making decisions about where to build a new health centre in Jeddah city. For example, areas beyond the 2Km accessibility zones can be used as a reference guide to enlist candidate locations for additional health centers in Jeddah city. The candidate locations can be presented to both regional and national health authorities for their consideration in building new health centers in selected parts of the city.

4.3. Health Centre-to-population Ratio

One way of defining accessibility to health centers is by identifying the provider-to-population ratios. This measure is useful for gross comparisons of supply between geo-political units or service areas, and it is used by policy analysts to set the minimum standard of local supply and to identify under served areas (Guagliardo et al. 2004). A GIS can be used to compute these ratios either through the density function as in Guagliardo et al. (2004) or the Buffer and overlay functions as in Luo (2004).

The ArcGIS spatial analyst is used in this study to create a GIS model that defines the provider-to-population ratio for health centers in Jeddah city in a three-step procedure.

(1) Provider density is created for each health centre

The density function in ArcGIS is useful for showing where point or line features are concentrated and can be calculated using the simple or Kernel methods. Based on the simple density method, points and lines that fall within the search area are summed and then divided by the size of the search area to obtain the density value for each cell. The Kernel density calculation is similar to that of simple density, except that points or lines nearer the centre of a raster/search cell are weighted more heavily than those along the edges of the search area to yield a smother surface of value distribution (McCoy and Jonhston 2001).

This study applied the Kernel function using several values for the search area before selecting 2000m as the most suitable search radius for

Jeddah city. Several parameters must be defined for the kernel function, such as input data source, density type and output cell size. In this case, the input to the density function was a point feature dataset representing the health centers. The ArcGIS software calculated the output cell size using the smaller of width or height of the feature extent and divided it by 250 to yield a cell size of 192.91m. The choice of a suitable raster cell size represents a major issue because the cell must be small enough to capture the required detail but large enough for processing efficiency.

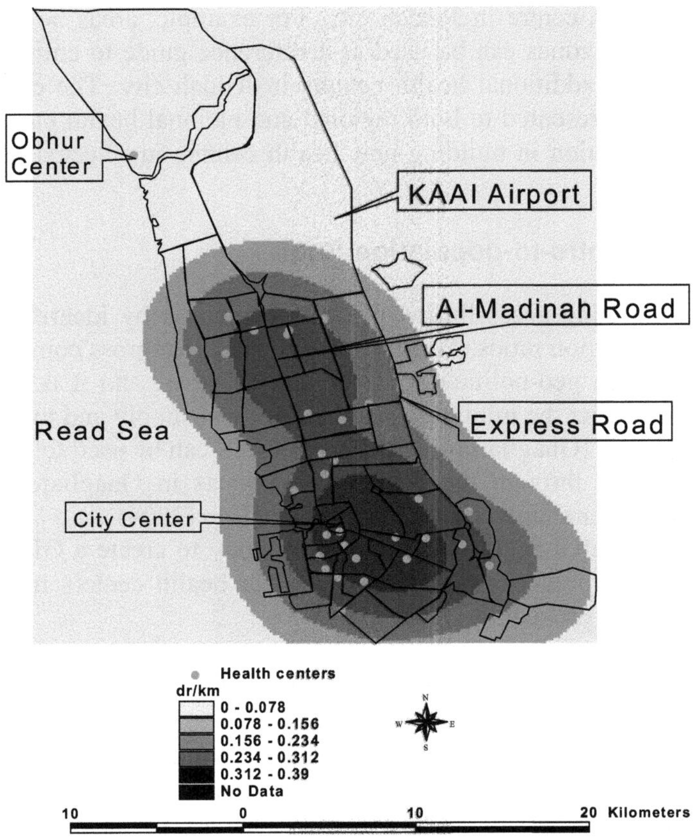

Fig. 2. Density of physicians per square kilometer of area

Figure 2 is a kernel map showing the density of physicians based on doctors available at every health centre. It is clear that two areas located closest to the city centre have a higher density of physicians. On the contrary, the northern areas have remarkably low ratios. This distribution pattern is attributed to two factors. First, if a health centre had a larger amount of doctors, then its surrounding areas would get a higher provider density. Second, the closer the health centers are to each other, the higher the provider density. This scenario was found mainly at the city centre where several health centers were located close to each other and, at the same time, these centers have a comparatively larger supply of doctors than those located north and west of the city.

(2) Population density is built for Jeddah city

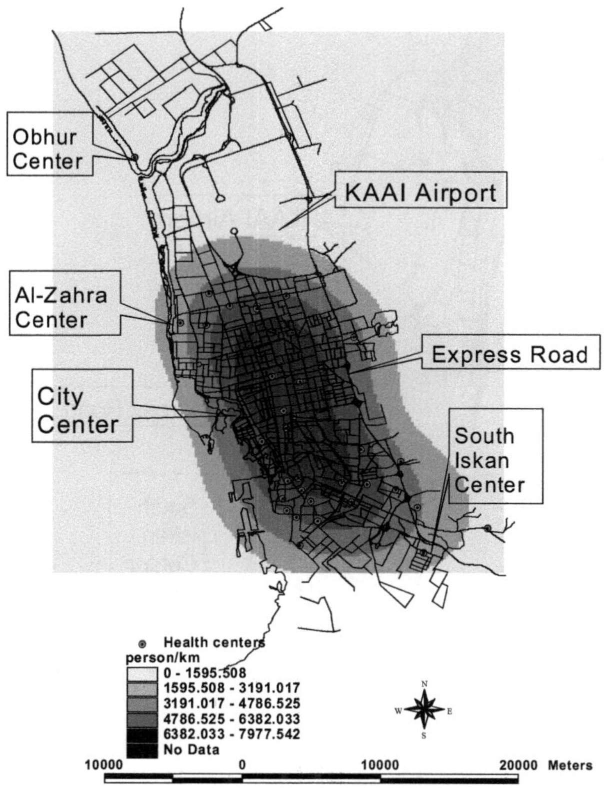

Fig. 3. A population density surface of the Jeddah city

The kernel density function was used again on the population data to produce a raster dataset for Jeddah city to describe the spatial variation of its population. The input data comprised point features that register a population size for every city districts. The search radius was 2000m and the output cell size was 214.65m.

This population density model was produced based on a GIS coverage of district centers (Figure 3). The spatial pattern of population density of Jeddah city indicates an increased density near the city centre and a decreased density in the northern city districts. This finding of a decreasing population density to the north of the city suggests that the demand for health services in the north is not as high as that in the central parts of the city.

<u>(3) Outputs from previous steps are analyzed</u>

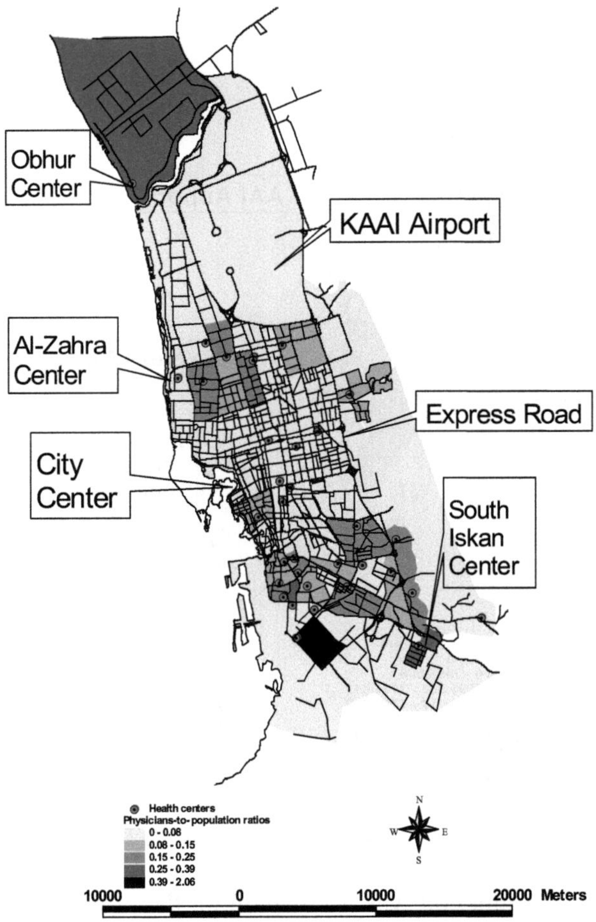

Fig. 4. Physician-to-population ratios in Jeddah city

Figures 2 and 3 produced earlier were analyzed further using the raster calculator function to perform mathematical operations. This type of GIS function was employed to assign weights to raster data for suitability analysis. The provider density (in physicians per square kilometer) was divided by the population density (in person per square kilometer). This step was achieved using the ArcGIS model builder that was based on the arithmetic overlay function. The output of this function is presented in Figure 4 which shows that the ratio is higher due north and south of the city districts. The remaining parts of Jeddah city have remarkably low ratios. This scenario reflects the imbalance health provision in the central part of the city with a larger population size but a limited supply of health centers.

The mathematical model are used by health planners to define which parts of Jeddah city are not well served by health centers and prepare accordingly new development plans for these areas. It is clear from this output that there are several parts of Jeddah city needing more supply of health facilities. These are located in the central and southern city districts. Should local health authorities have plans to increase the supply of health facilities, then this model offers useful insights on the matter.

4.4. Modeling Demand Flows

The gravity model (in general) and spatial interaction (in particular) have been used as valid measures of spatial accessibility. These models combine indicators of distance and availability of health supply using the following formula:

$$Pij = Ai \times Ri \times Dj \times f(Cij) \tag{1}$$

where Pij = flow of patients from residential area i to health centre j,
Ri = patient demand in area I,
Dj = number of physicians at health centre j,
Cij = is a measure of the cost of travel or distance between i and j, and
Ai = is a balancing factor.

The study applied the above formula for Jeddah city using the ArcGIS Network Analyst module. The output of this model is an Info table showing the spatial interaction between every health centers and their areas of demand. The table shows results based on three types of constraints which are: Production-constrained interaction, Attraction-constrained interaction, and Doubly constrained interaction. Each one of these constrained models computes interaction based on some types of restraint to balance the results of the known inputs (ESRI 1991).

North Obhur and South Iskan were two health centers selected to present results of the Info table. Both results (Figures 5a and 5b) show the decreasing zones of interaction from the selected centers due to distance decay and proximity to competitors. These results are important for health planners because they serve to visualize how well existing health centers serve different areas and target groups, as well as to identify where important gaps have remained (Birkin, et al. 1996). Looking at these results, it can be said that there are more gaps in the catchment area of the South Iskan health centre than of the North Obhur centre.

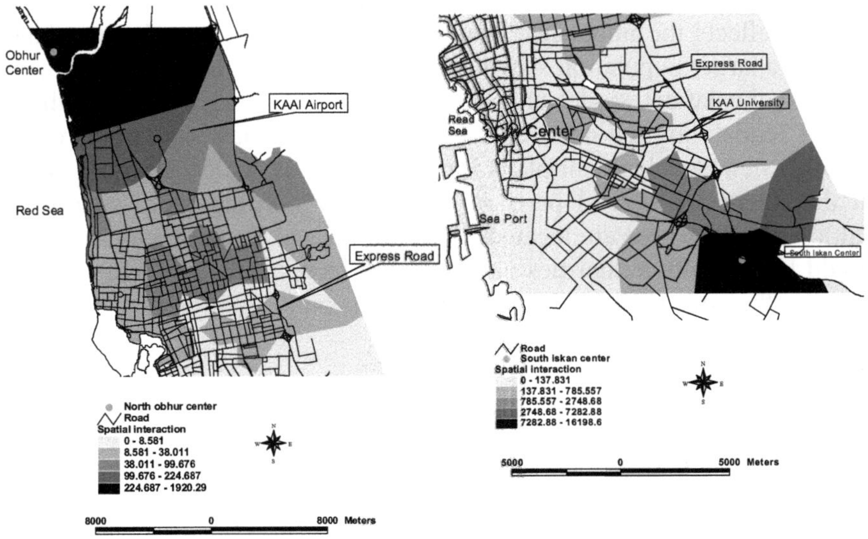

(a) North Obhur health centre (b) South Iskan health centre

Fig. 5. Predicted demand flows

Areas around the South Iskan centre have a low health accessibility, and therefore, an increase in health service provision is recommended for that centre. However, the situation is slightly different at the North Obhur centre because areas around that centre show a wider health catchment area.

These kinds of interaction models are very useful for allocating flows of patients between origins and destinations on the basis that flows between origins and destinations are proportional to the relative attractiveness and accessibility of the destinations (Murad 2004). The two main usages of spatial interaction models include defining the variation of health demand flows and to evaluating the performance of health services. Looking at the predicted demand flows of the North Obhur health centre, for example, we can observe that the level of demand flow decreases when distance to the health centre increases. This pattern is also exhibited by the south Iskan health centre where many areas located far from the centre show a remarkably low health demand.

5. Conclusion

The study employed several GIS functions and models for the purpose of defining spatial accessibility to health centers in Jeddah city. Different data sets were collected and captured using the ArcGIS software. The created database was used to build three spatial accessibility models which provided Health planners of Jeddah city options to evaluate the current status of health centers and their spatial locations. One major finding from the accessibility models is that there is a need to increase the current supply of health centers in Jeddah city. A better health care service will surely be achieved in Jeddah city were an increase in health services implemented.

6. References

[1] Al-Sahili K and Abdul-Ella M (1992) Accessibility of Public Services in Irbid, Jordan. Journal of Urban Planning and Development 1180: 1-12.

[2] Birkin M, Clarke G, Clarke M, Wilson A (1996) Intelligent GIS: Location decisions and strategic planning. Geo information, Cambridge

[3] Grimshaw D (2000) Bringing geographical information systems into Business. John Wiley & Sons, New York

[4] Guagliardo M, Ronzio C Cheung I, Chacko E, Joseph J (2003) Physician accessibility: an urban case study of pediatric providers. Health & Place 10: 273-283

[5] Jong T and Van Eck R (1997) Location Profile-based measures as an improvement on accessibility modeling in GIS. Computer Environment and Urban systems 2: 181-190

[6] Luo W (2004) Using a GIS-based floating catchment method to assess areas with shortage of physicians. Health & Place 10: 1-11

[7] McCoy J and Johnston K (2001) Using ArcGIS spatial analyst. ESRI, Redlands

[8] Moseley M (1979) Accessibility: the rural challenge. Methuen & Co ltd, London

[9] Murad (2004) Creating A GIS Application for Local Health Care Planning in Saudi Arabia. International Journal of Environmental Health Research 14: 185-199.

[10] Phillips D (1990) Health and health care in the third world. Longman, Essex.

[11] Thouez J, Bodson P, Joseph A (1988) Some methods for measuring the geographic accessibility of medical service in rural regions. Medical Care 26: 34-44

[12] Wang F and Luo W (2004) Assessing spatial and no spatial factors for healthcare access: towards an integrated approach to defining health professional shortage areas. Health & Place11: 131-146

Human and Environmental Factors

Applying GIS in Physical Activity Research: Community 'Walkability' and Walking Behaviors

Ester Cerin[a], Eva Leslie[b], Neville Owen[c] and Adrian Bauman[d]

[a] Institute of Human Performance, University of Hong Kong, Hong Kong SAR, China
[b] School of Health and Social Development, Deakin University, Australia
[c] Cancer Prevention Research Centre, University of Queensland, Australia
[d] NSW Centre for Physical Activity and Health, University of Sydney, Australia

---

Abstract: Physical activity provides many health benefits, including reduced risk of cardiovascular disease, Type II diabetes and some cancers. Environmental exposure factors (e.g., the built environment) are now receiving ever-increasing attention. Large-scale interdisciplinary studies on the association between attributes of local community environments and residents' physical activity are being conducted. We will focus on findings from Australia - the Physical Activity in Localities and Community Environments (PLACE) study. PLACE is examining factors that may influence the prevalence and the social and spatial distribution of walking for transport and walking for recreation. A stratified two-stage cluster sampling strategy was used to select 32 urban communities (154 census collection districts), classified as high and low 'walkable' using a GIS-based walkability index (dwelling density, intersection density, net retail area and land use mix) and matched for socio-economic status. We report data on a sub-sample of 1,216 residents who provided information on the perceived attributes of their community environments (e.g., dwelling density, access to services, street connectivity) and weekly minutes of walking for transport and for recreation. Moderate to strong associations were found between GIS indicators of walkability and the corresponding self-report measures. The walkability index explained the same amount of neighborhood-level variance in walking for transport as did the complete set of self-report measures. No significant associations were found with walking for recreation. Relevant GIS-based indices of walkability, for purposes other than transport need to be developed.

Keywords: GIS, community walkability, walking behavior, Australia, adults.

1 Introduction

Regular physical activity has been shown to be a protective factor for various chronic diseases, including cardiovascular disease, Type II diabetes, osteoporosis and some cancers (USDHHS 1996). Walking is the most common form of health-enhancing physical activity of adults (Hayashi et al. 1999), and makes up a significant proportion of their total physical activity (Rafferty et al. 2002).

Strategies to promote physical activity have commonly taken a person-focused approach with interventions developed to target cognitive and social determinants (King et al. 2002). However, the built environment can also act as a determinant of physical activity behavior (Sallis and Owen 2002). Empirical support for a significant impact of the built environment on walking has been accumulating in two different disciplines. Within the transportation and urban planning literature, a recent review of literature has identified 14 studies that examined and found an association between neighborhood physical attributes and residents' rates of engagement in active modes of transport (cycling and walking) (Saelens et al. 2003a). Residents from neighborhoods with higher levels of residential density, street connectivity, and land use mix reported more walking and cycling than their counterparts. Similarly, within the health and behavioral science literature, Owen and colleagues (2004) examined 18 studies and found that aesthetic attributes, convenience of facilities for walking (sidewalks, trails), accessibility of destinations and perceptions about traffic and busy roads were associated with walking for particular purposes. They also found that attributes associated with walking for exercise were different from those associated with walking for transport.

Despite these encouraging preliminary findings, many questions on the environment-physical activity relationships and how they interact with socio-demographic factors remained to be answered. An accurate analysis of these relationships implies the use of valid and accurate measures of physical activity as well as accurate measures of attributes of the neighborhood built environment. Attributes of the built environment can be measured objectively (e.g., using Geographic Information Systems) and subjectively (e.g., using questionnaires). Both types of measures are important because walking and other types of physical activity behavior are likely to depend on the objective environment as well as on how a person interprets the environment (Owen et al. 2004).

One of the main aims of this chapter was to examine the correspondence between an objective, GIS-based measure of community walkability (walkability index) (Leslie et al. 2007) and perceived attributes of the environment believed to be associated with walking for transport and recreation. The second main aim was to examine the association of these measures with self-reported walking for transport and walking for recreation. In doing this, we used a subset of data from the Physical Activity in Localities and Community Environments (PLACE) study conducted in Adelaide (Australia).

2 Methods

2.1 Study Design and Participants

The PLACE study was based on the Neighborhood Quality of Life Study conducted by Sallis and colleagues in the USA (www.nqls.org). The aim of both studies was to investigate associations between the built local community environments and residents' physical activity. Their study design has been adopted and is being promoted by the International Physical Activity and the Environment Network (www.ipenproject.org), whose goals are to support researchers investigating environmental correlates of physical activity by recommending common methods and measures and by stimulating collaboration and communication among researchers.

A stratified two-stage cluster sampling design was used to recruit participants. Firstly, 32 communities were selected from 2078 urban census collection districts[2] (CCDs) in the Adelaide Statistical Division. The communities comprised contiguous clusters of CCDs that were identified as high or low walkable using a GIS-based walkability index (see Figure 1) and matched for socio-economic status (using census data on median household weekly income and data on property valuations). Secondly, households were randomly drawn (without replacement) from a list of residential addresses within the 32 selected communities. Eligible respondents were adults, aged 20-65, able to walk without assistance, able to take part in surveys in English, and residing in private dwellings such as houses, flats or units. In households with more than one eligible participant, the person with the most recent birthday was asked to take part in the

[2] A CCD is the smallest administrative unit used by the Australian Bureau of Statistics (ABS) to collect census data.

study. Participant recruitment and data collection was by mail. Participants (N = 2,652) completed a survey and mailed it back to the research team. Thank-you letters and a lottery based incentive were provided to participants on return of their completed questionnaire. The study was approved by the Behavioral and Social Sciences Ethics Committee of the University of Queensland. For a more detailed account of the recruitment methods see du Toit et al. (2005).

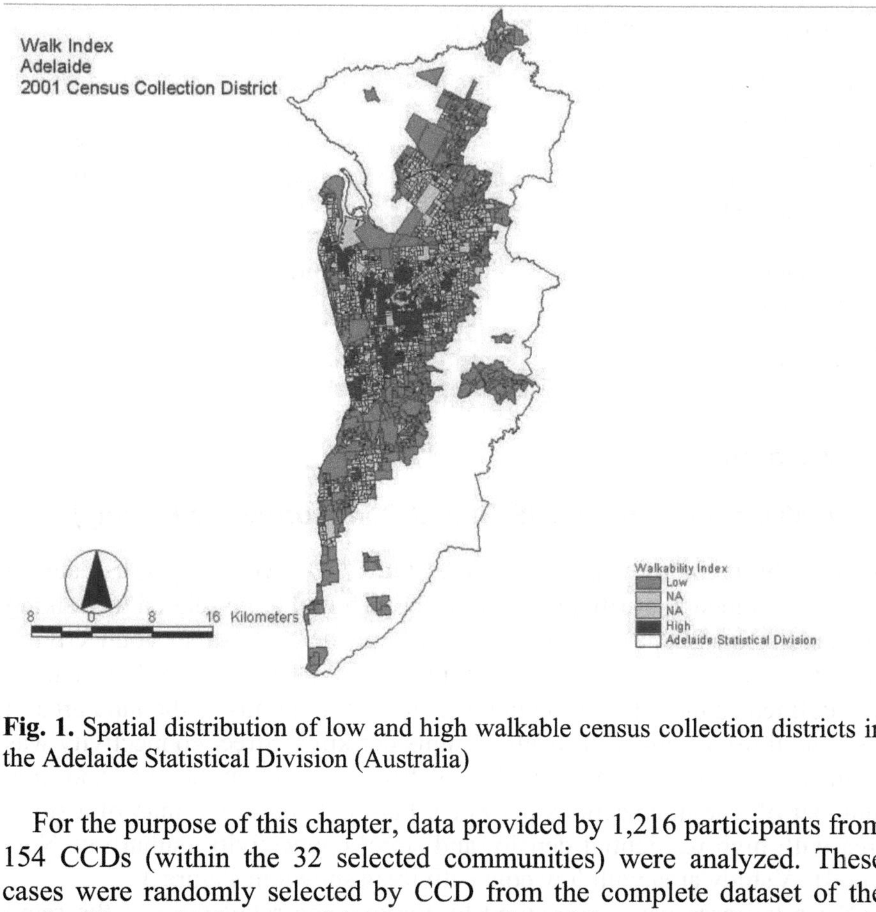

Fig. 1. Spatial distribution of low and high walkable census collection districts in the Adelaide Statistical Division (Australia)

For the purpose of this chapter, data provided by 1,216 participants from 154 CCDs (within the 32 selected communities) were analyzed. These cases were randomly selected by CCD from the complete dataset of the study. The socio-demographic characteristics of the examined sub-sample are summarized in Table 1.

Table 1. Socio-demographic characteristics of the sample (n = 1,216)

Characteristic	Estimate	Characteristic	Estimate
Male, %	38.2	Education, %	
Missing values	0.5	Year 10 or less	21.0
Employment status, %		Year 12 / trade	28.5
Full time	39.5	Tertiary	46.8
Part time / casual / family	26.1	Missing values	3.7
Unemployed or home duties	13.6	Annual household income, %	
Retired / permanently unable to work	13.7	< $20,800	22.0
Other	5.1	$20,800 - $41,599	25.8
Missing values	1.9	$41,600 - $77,999	29.2
Age, mean (SD), y	42.8 (11.5)	> $77,999	18.2
Missing values, %	1.0	Missing values	4.8
Children in household, %			
Yes	32.1		
Missing values	1.0		

2.2 Measures

2.2.1 Walkability Index (GIS-Based Community Walkability)

A Walkability Index was computed at the CCD level using GIS data on four environmental attributes that have been found to be related to walking (Saelens et al. 2003a). These were dwelling density, street connectivity, net retail area and land-use mix (Frank and Pivo 1994). Dwelling density was defined as the ratio of the number of dwelling units to the land area in residential use (number of dwelling units per km^2). Street connectivity was defined as the ratio of the number of true intersections to the land area (number of true intersections per km^2). Figure 2 shows the map of a study area with high residential density and street connectivity (suburb of Norwood). A study area with low connectivity is shown in Figure 3.

Net retail area was represented by the average retail ratio of the retail gross floor space to the land area. Finally, land-use mix represented the degree to which different land uses were scattered within the land area. Land-use mix was quantified as an entropy index. Land use was classified into the following categories: residential, commercial, industrial, recreational and other. The entropy index was computed using the following formula (Frank and Pivo 1994):

$$-\frac{\sum_k (p_k \ln p_k)}{\ln N} \qquad \text{Eqn. (1)}$$

where k is the category of land use; p is the proportion of the land area within a CCD allocated to a specific land use; and N is the number of land use categories (in our case, five). The entropy index ranges from 0 to 1, with 0 representing homogeneity (all land uses are of a single type), and 1 representing maximal heterogeneity (the land use categories are evenly represented in the CCD). The four environmental characteristics were transformed into deciles and summed for each CCD, with a possible score of 4 to 40, denoting the Walkability Index of a CCD (see Figure 1). Similar walkability indexes have been related to physical activity variables in studies conducted in the Atlanta, Georgia (Frank et al. 2005) and Seattle, Washington regions (Frank et al. 2006), supporting its validity and generalizability.

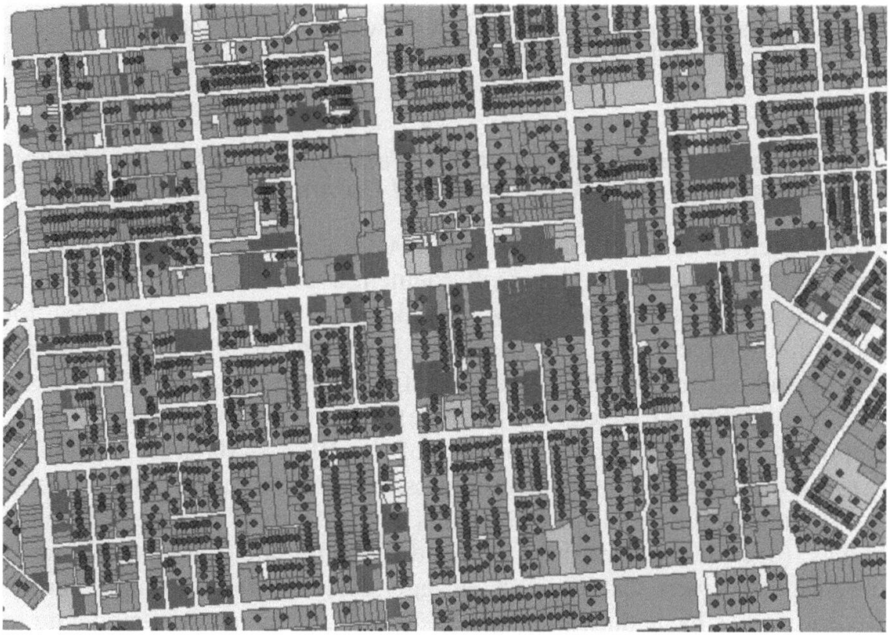

Fig. 2. Map section representing a high density and high street connectivity study area (suburb of Norwood, Adelaide, Australia); dots represent parcels in residential use

Fig. 3. Map section representing a study area with low street connectivity (Salisbury Heights, Adelaide, Australia)

2.2.2 Neighborhood Environment Walkability Scale – Australian Version (NEWS-AU; Perceived Community Walkability)

This 63-item instrument (Saelens et al. 2003b; Leslie et al. 2005) measured perceived attributes of the local environment hypothesized to be related to physical activity and, particularly, to walking for transport and walking for recreation. Concepts and subscales were based on variables believed to relate to walking and other physical activities that are discussed in the urban planning literature (Saelens et al. 2003a). For the purpose of this chapter, responses on the following subscales of the NEWS-AU were examined: (a) ease of access to non-residential uses; (b) street connectivity; (c) walking/cycling facilities, such as sidewalks and pedestrian/bike trails; (d) aesthetics; (e) pedestrian traffic safety; and (f) crime safety. Items were rated on a 4-point Likert scale from 1 (strongly disagree) to 4 (strongly agree). Recent studies have shown that the NEWS-AU items possess moderate to high levels of test-retest reliability (Leslie et al. 2005).

2.2.3 Walking for Transport and Walking for Recreation

Weekly minutes of walking for transport and walking for recreation were assessed using relevant items from the long version of the International

Physical Activity Questionnaire (IPAQ; long last-7-days self-administered format; Craig et al. 2003). A recent report on data collected in 12 countries showed that the IPAQ had comparable reliability and validity to other self-report measures of physical activity (Craig et al. 2003). The respondents were instructed to report the frequency and duration of walking for transport and walking for recreation during the past seven days. These items of the IPAQ have been shown to possess good reliability (Vandelanotte et al. 2005) and adequate validity (Vandelanotte et al. 2005; Cerin et al. 2006). In this study, total weekly minutes of walking for transport and for recreation were computed and values were truncated to 1860 minutes.

2.2.4 Socio-Demographic Attributes

Participants were asked to report their age, gender, educational attainment, annual household income before taxes, employment status and number of children under 18 years in the household.

2.3 Data Analyses

Multilevel confirmatory factor analysis (MCFA) was employed to define a neighborhood-level measurement model for the NEWS-AU subscales with items rated on a 4-point Likert scale. We used MCFA because this study adopted a two-stage cluster sampling design and significant intraclass correlations (ICCs), denoting the proportion of total item variance due to differences between CCDs, were observed for all the NEWS-AU items (Muthén 1997). MCFA was carried out using Bentler and Liang's Maximum Likelihood Estimation (MLE) method (Bentler and Liang 2003). This method uses an EM-type gradient algorithm for computing the MLE for two-level structural equation models, of which confirmatory factor analysis is a special case. This algorithm is applicable to any sample size with balanced or unbalanced design and is preferable to other methods when the sample sizes vary substantially among clusters (neighborhoods). The end result of a MCFA is the estimation of a measurement model for each level of variation (within- and between-neighborhoods). This chapter focuses on between-neighborhood variations and, hence, a neighborhood-level measurement model of the NEWS.

The construction of a two-level measurement model of the NEWS included three main steps. The first two steps involved the assessment of two *a priori* two-level measurement models. The first model encompassed six oblique (correlated) factors, as originally defined by its developers (Saelens et al. 2003b). For this model, the factor structures at the individual and neighborhood level were defined to be equal. The second *a priori* model consisted of five correlated neighborhood-level factors, and six correlated individual-level factors, emulating the empirically-based measurement model of the original version of the NEWS (Cerin et al. 2006). In the third step of the analyses, re-specification of the original model was conducted according to Jöreskog and Sörbom's (1993) iterative model-generating approach. This approach consists of testing the viability of initial hypothetical models in terms of whether they satisfactorily fit the observed data. If the results indicate a lack of fit based on empirical or substantive evidence, the models are re-specified. The ultimate goal of model re-specification is to identify models that can provide a statistically acceptable fit and a theoretically meaningful interpretation of the data. Model re-specification was based on the analysis of standardized factor loadings, the analysis of three empirical indices of poor model fit (standardized residual covariances, univariate Lagrange multiplier tests and Wald tests), and substantive considerations (i.e., salience) (15). Factor loadings greater than |.30| were considered to be significant (Bryant and Yarnold 1994).

Several measures of absolute and incremental fit were used to evaluate the goodness-of-fit of the measurement models. Absolute-fit indices describe the ability of the model to reproduce the original covariance matrix. The absolute-fit indices reported in this chapter are the goodness-of-fit index (GFI) and the root mean square error of approximation (RMSEA). The GFI describes the proportion of variance in the original data that is accounted for by the proposed model, with values ranging from 0 (*poor fit*) to 1.0 (*perfect fit*). The RMSEA estimates the average covariance residual that would have been expected had the entire population of potential respondents been represented, with values less than .08 indicating an acceptable model fit. Two incremental fit indices, assessing the degree to which a specified model is better than a baseline model that specifies no covariances, were used. These were the non-normed fit index (NNFI) and the comparative fit index (CFI). Values exceeding .90 are generally indicative of an acceptable model fit. In addition to the above indices of fit, the standardized root mean squared residual (SRMR) was computed. The SRMR is a standardized summary of the average covariance residuals. A favorable value for the SRMR is less than .10 (Browne and Cudeck 1993). All MCFA analyses were performed using EQS 6.1 (Multivariate Software Inc. 2004).

Multilevel linear models were used to estimate the magnitude of the neighborhood-level associations between the GIS-based walkability index/components (neighborhood-level variables) and the MCFA-based neighborhood-level NEWS-AU subscales. These models were also used to estimate the neighborhood-level associations of the walkability index, its components and the NEWS-AU factors with weekly minutes of walking for transport and walking for recreation. These associations were estimated following the procedure specified by Snijder and Bosker (1993). A probability level of 0.05 was adopted. These analyses were performed using MLwiN 2.02 (Multilevel Models Project, Institute of Education 2004).

3 Results

The two *a priori* models of the NEWS-AU did not adequately fit the observed data. Hence, these models were re-specified following the procedure described above. Table 2 reports the final between-neighborhood measurement model (groups of items and underlying factors) of the NEWS-AU. Three latent factors of perceived neighborhood environmental attributes were found. All factor loadings were significant at a probability level of .05 and their absolute values were greater than .42. These were 'Access to services and ease of walking', 'Street connectivity' and 'Aesthetics, traffic and safety'. The first two factors were moderately positively correlated (.48). All the goodness-of-fit indices indicated an acceptable fit of the final measurement model to the data (GFI = .96; RMSEA = .052; SRMR = .100; NNFI = .91; CFI = .90).

The descriptive statistics of, and correlations between, the walkability index, its components and the NEWS-AU factors are shown in Table 3. A reasonable degree of variability was observed in the objective and perceived measures of neighborhood environmental attributes (see means and standard deviations in Table 3). Ratings on the NEWS-AU factors of 'Access to services and ease of walking' and 'Street connectivity' were moderately to strongly correlated with the walkability index and its components ($r = 0.32$ to 0.66), with the exception of land-use mix. No significant associations were observed between the NEWS-AU factor 'Aesthetics, traffic and crime safety' and the GIS-based measures of neighborhood walkability.

Table 2. Neighborhood-level measurement model of the NEWS-AU

Neighborhood-level factor	Item content
Access to services & ease of walking	- Can do most of the shopping - Many shops within easy walking distance - Many places to go within easy walking distance - Easy to walk to a public transport stop - Major barriers to walking [negatively scored] - Footpaths on most of the streets - Well-maintained footpaths - Busy streets have pedestrian crossing / traffic signals
Street connectivity	- Streets do not have many, or any, cul-de-sacs - Walkways that connect cul-de-sacs to streets [negatively scored] - Short distance between intersections. - Many four-way intersections - Many alternative routes for getting from place to place - Footpaths separated from the road/traffic by parked cars
Aesthetics, traffic & crime safety	- Park or nature reserve in the local area - Bicycle or walking paths in or near the local area - Lots of greenery around the local area. - Tree cover or canopy along the footpaths - Many interesting things to look at while walking - Local area free from litter, rubbish, or graffiti - Many attractive buildings and homes - Pleasant natural features - A lot of traffic along the streets [negatively scored] - Main arterial road or busy throughway for motor vehicles nearby [negatively scored] - A lot of exhaust fumes [negatively scored] - A lot of petty crime [negatively scored] - A lot of major crime [negatively scored] - Level of crime makes it unsafe to walk during the day [negatively scored] - Level of crime makes it unsafe to walk at night [negatively scored] - Feel safe walking home from a bus or train stop at night

Table 3. Neighborhood-level correlations between perceived and objective attributes of the neighborhood environment (154 CCDs)

		GIS measures				
		WI	*DD*	*ID*	*NRA*	*LUM*
NEWS factors	*M* (*SD*)	18.9 (5.5)	720.7 (304.3)	245.5 (156.4)	0.24 (0.20)	0.41 (0.19)
ASEW [a]	3.25 (0.64)	0.48*	0.32*	0.38*	0.48*	0.08
SC [a]	2.94 (0.63)	0.69*	0.48*	0.55*	0.65*	0.09
ATCS [a]	2.75 (0.74)	-0.10	0.01	0.06	-0.10	0.02

M = mean; *SD* = standard deviation; *ASWE* = Access to service and ease of walking; *SC* = Street connectivity; *ATCS* = Aesthetics, traffic and crime safety; *WI* = walkability index (possible range of values 4-40); *DD* = dwelling density (defined as dwelling units per km^2); *ID* = intersection density (defined as number of true intersections per km^2); *NRA* = net retail area (defined as the ratio of the gross retail area to the total retail parcel area); *LUM* = land-use mix (possible range of values 0 – 1;); [a] possible range of values 1-4; * $p < .05$

Table 4. Neighborhood-level correlations of perceived and objective attributes of the neighborhood environment with self-reported weekly minutes of walking for transport

Objective measures	*R*	Perceived measures	*R*
Walkability index	.20*	Access to services and ease of walking	.27*
Dwelling density	.45*	Street connectivity	.41*
Intersection density	.27*	Aesthetics, traffic and crime safety	-.18
Net retail area	.18		
Land-use mix	-.10		

R = multiple correlation; * $p < .05$

Table 4 reports the univariate neighborhood-level associations of objective and perceived measures of neighborhood walkability with walking for transport, adjusting for individual-level socio-economic factors. As no significant neighborhood-level variation in walking for recreation was observed, neighborhood-level associations between this type of walking and objective/perceived attributes of the environment could not be computed. The examined sample (N = 1,216) walked for transport on average 180 minutes a week (*SD* = 279 min/wk; Median = 85 min/week). Neighborhood differences explained 1.4% of the variations in residents' walking for transport.

Significant positive associations were found between weekly minutes of walking for transport and the GIS-based walkability index, dwelling density, and intersection density (Table 4). Although the relationship between net retail area and walking for transport was in the expected direction, it was not statistically significant. Also, the data did not provide support for an association between the GIS-based measure of land-use mix and walking for transport. As expected, two neighborhood-level factors of the NEWS-AU hypothesized to be related to utilitarian walking were indeed significantly related to weekly minutes of walking for transport. In contrast, the factor 'Aesthetics, traffic and crime safety' was not a reliable explanatory variable of self-reported walking for transport. The multiple correlation between the four GIS-based, objective measures of walkability and walking was 0.47, whilst that between the three survey-based measures of perceived walkability and walking was 0.45.

4 Discussion

One of the main aims of this chapter was to examine the correspondence between GIS-based, objective measures and survey-based, perceived measures of neighborhood walkability in a sample of Australian adults. Moderate to high levels of correspondence were found between perceived neighborhood access to services, ease of walking, street connectivity and all but one of the GIS-based measures of walkability. Specifically, no significant association was observed between these perceived aspects of the neighborhood environment and objective land-use mix. It is possible that the lack of correspondence was due to the way land-use mix was operationalized. The measure of land-use mix used in this study was based on the distribution of land area across different uses rather than, as recommended, on the distribution of building floor space across different uses. This can result in very different outcomes as, for example, a tall building may contain over a million square feet of office and commercial use on a 1 acre (43,560 square foot site). However, adequate building square foot data were not available for Adelaide (Australia). Also, while previous studies used a measure of land-use mix which was exclusively based on categories of land use that are thought to be conducive to walking for transport (e.g., commercial, residential, institutional, office; Frank et al. 2004; Frank et al. 2005), this study included an additional category of land use – namely, recreational land use. Access to this type of land use is not captured by the NEWS-AU factors of 'Access to services and ease of walking' and 'Street connectivity'. This could at least in part explain why no significant association was found between a GIS-based measure of objective land-use mix and perceived attributes of the neighborhood environment thought to be related to utilitarian walking.

The fact that the NEWS-AU factors 'Access to services and ease of walking' and 'Street connectivity' were, but 'Aesthetics, traffic and crime safety' was not related to the objective measures of walkability, gives support for the construct and criterion validity of the NEWS-AU as a measure of neighborhood environmental attributes. As the GIS-based measures of walkability were essentially measures of access to services and street connectivity, it was logical to expect that these would show a significant association with their corresponding perceived neighborhood attributes (NEWS-AU factors). Moreover, since the GIS-based measures of walkability did not include information about aesthetic aspects of the neighborhood environment, green areas, traffic load and safety from crime, it is not surprising that no relationship was observed between the GIS-based walkability measure and the respondents' ratings on the factor 'Aesthetics, traffic and crime safety'.

The second main aim of this study was to examine the relationships of objective and perceived attributes of the neighborhood environment with walking for transport and walking for recreation. The associations between the GIS-measures of walkability and walking for transport were, in the main, significant and in the expected direction (i.e., positive). Contrary to previous studies (Frank et al. 2004), no significant association was found between objective land-use mix and walking for transport. As noted earlier, this finding may be attributable to how land-use mix was operationalized. The index of land-use mix used in this study encompassed a component (i.e., recreational land use) that is more likely to be associated with walking for recreation. In this regards, Cerin et al. (in press) found that residents of areas with access to commercial and industrial destinations reported significantly more transport-related walking than did residents of areas with access to recreational destinations, even if these areas had a similar level of land-use mix. Indeed, recreational venues are likely to be regularly visited less often and by fewer people (Sallis et al. 2004) and, hence, their impact on residents' transport-related walking is bound to be limited.

Positive associations were also observed between some of the perceived neighborhood attributes and walking for transport. As was the case for their objective counterparts, perceived access to services and street connectivity in the neighborhood of residence were predictive of higher levels of walking for transport. Similar results have been recently reported in the transport and urban planning literature and public health literature (Transportation Research Board 2005). No significant associations were found between walking for transport and perceived aesthetic features of the neighborhood environment, traffic and safety from crime. This is somewhat understandable as these are characteristics that haven been shown to be more pertinent in relation to walking for recreation (Owen et al. 2004).

Importantly, in the present investigations, the GIS-based measure of walkability was as good an explanatory variable of neighborhood differences in walking for transport as were residents' perceptions of the neighborhood environment. Both measures accounted for a significant proportion of the neighborhood-level variance in walking for transport (i.e., over 20%). This is an important finding supporting the utility of GIS for the measurement of physical activity correlates and determinants. Additionally, this particular finding suggest that the strength of association between attributes of the built environment and walking behavior may be substantially greater than previously noted. Studies examining the environment-walking association at the individual level have reported only up to 5% of shared variance between walking measures and specific aspects of the environment (Transportation Research Board, 2005). These weaker associations can be partly attributed to the fact that some individual differences in walking and perceived neighborhood environment are due to measurement error (Cerin et al. 2006). An examination of the environment-walking association at the neighborhood level can partly eliminate such problems. This is because the average rating given by a group of residents for the same neighborhood is likely to be a more reliable and valid measure of the environment than is the rating of a single individual.

Finally, no significant association could be found between both objective and perceived neighborhood environmental attributes and walking for recreation. This is because there was no neighborhood-level variation in the respondents' level of self-reported walking for recreation. This finding can be easily explained by noting that, in this study, neighborhood selection was based on attributes that are relevant to walking for transport (e.g., street connectivity, dwelling density, and access to services) but irrelevant to walking for recreation. Future research needs to focus on the development of a GIS-based index of walkability suited to walking for recreation. This index might include information on green areas, traffic load, presence of footpaths, and proximity to beach or waterways, as suggested by recent studies in the area of public health (Humpel et al. 2003).

5 Conclusions

In general, this study provides further support for the contention that aspects of the built environment are associated with specific forms of physical activity, and, hence, residents' health. We have illustrated how GIS data were successfully used in the PLACE study to define and measure aspects of the built environment that are related to walking for transport. Importantly, in the present investigation, a GIS-based measure of walkability, grounded on the work of Frank and Pivo (1994), emerged to be as good an explanatory variable of neighborhood differences in walking for transport as residents' perceptions of the neighborhood environment. This confirms the utility of using GIS in physical activity research. This chapter has also highlighted that there is still room for improvement in the operationalization of GIS-based measures of neighborhood transport-related walkability. There is also a need for the development of a walkability index appropriate for walking for recreation.

Acknowledgements

The National Health and Medical Research Council (NHMRC) Project Grant #213114 and NHMRC Program Grant #301200 supported the PLACE (Physical Activity in Localities and Community Environments) study. The authors are grateful to the South Australian Government department for Transport and Urban Planning for providing access to the relevant GIS data used in this study. We thank Mr. Neil T. Coffee (National Centre for the Teaching and Research into Social Application of GIS, University of Adelaide) for computing the GIS-based walkability index and producing the maps presented in this chapter. We also thank Dr. Poh-Chin Lai and the GIS laboratory of the Department of Geography, University of Hong Kong for processing the images included in this chapter.

References

[1] Bentler PM and Liang J (2003) Two-level mean and covariance structures: maximum likelihood via an EM algorithm. In: Reise SP, Duan N (eds.) Multilevel modeling: Methodological advances, issues, and applications. Lawrence Erlbaum, Mahwah NJ, pp 53-70

[2] Browne MW and Cudeck R (1993) Alternative ways of assessing model fit. In: Bolen KA, Long JS (eds.) Testing structural equation models. Sage Publications, Newbury Park CA, pp 136-162

[3] Bryant FB and Yarnold PR (1994) Principal-components analysis and exploratory and confirmatory factor analysis. In: Grimm LG, Yarnold PR (eds.) Reading and understanding multivariate statistics. APA, Washington DC pp 99-136

[4] Cerin E, Leslie E, du Toit L, Owen N, Frank L (in press) Destinations that matter: associations with walking for transport. Place Health

[5] Cerin E, Saelens BE, Sallis JF, Frank LD (2006) Neighborhood Environment Walkability Scale: validity and development of a short form. Med Sci Sports Exerc 38:1682-1691
[6] Craig CL, Marshall AL, Sjöström M, Bauman A, Booth ML, Ainsworth BE, Pratt M, Ekelund U, Yngve A, Sallis J, Oja P (2003) International Physical Activity Questionnaire (IPAQ): 12-country reliability and validity. Med Sci Sports Exerc 35:1381-1395
[7] du Toit L, Cerin E, Leslie E (2005) An account of spatially based survey methods and recruitment outcomes of the Physical Activity in Localities and Community Environments (PLACE) study. Cancer Prevention Research Centre, School of Population Health, The University of Queensland, Brisbane Australia
[8] Frank L and Pivo G (1994) Impacts of mixed use and density on utilization of three modes of travel: single-occupant vehicle, transit, and walking, Transport Res Rec 1466:44-52
[9] Frank L, Andresen MA, Schmid TL (2004) Obesity relationships with community design, physical activity, and time spent in cars. Am J Prev Med 27:87-96
[10] Frank LD, Sallis JF, Conway T, Chapman J, Saelens B, Bachman W (2006) Many pathways from land use to health: walkability associations with active transportation, body mass index, and air quality. J Am Plann Assoc 72:75-87.
[11] Frank LD, Schmid TL, Sallis JF, Chapman J, Saelens BE (2005) Linking objectively measured physical activity with objectively measured urban form: findings from SMARTRAQ. Am J Prev Med 28(2S2):117-125
[12] Hayashi T, Tsumura K, Suematsu C, Okada K, Fujii S, Endo G (1999) Walking to work and the risk for hypertension in men: The Osaka health survey. Ann Intern Med 131:21-26
[13] Humpel N, Owen N, Leslie E, Marshall AL, Bauman AE, Sallis JF (2003) Associations of location and perceived environmental attributes with walking in neighborhoods. Am J Health Promot 18:239-242
[14] Jöreskog KG and Sörbom D (1993) Structural equation modeling with the SIMPLIS command language. SPSS, Chicago IL
[15] King AC, Stokols D, Talen E, Brassington GS (2002) Theoretical approaches to the promotion of physical activity. Am J Prev Med 23(S):15-25
[16] Leslie E, Coffee N, Frank L, Owen N, Bauman A, Hugo G (2007) Walkability of local communities: Using Geographic Information Systems to objectively assess relevant environmental attributes. Health Place 13:111-122
[17] Leslie E, Saelens B, Frank L, Owen N, Bauman A, Coffee N, Hugo G (2005) Residents' perceptions of walkability attributes in objectively different neighborhoods: a pilot study. Health Place 11:227-236
[18] Muthén BO (1997) Latent variable modeling of longitudinal and multilevel data. In: Raftery AE (ed.) Sociological methodology (pp. 453-481) ASA, Washington DC, pp 453-481
[19] Owen N, Humpel N, Leslie E, Bauman A, Sallis JF (2004) Understanding environmental influences on walking: review and research agenda. Am J Prev Med 27:67-76

[20] Rafferty AP, Reeves MJ, McGhee HB (2002) Physical activity patterns among walkers and compliance with public health recommendations. Med Sci Sports Exerc 34:1255-1261
[21] Saelens BE, Sallis JF, Frank LD (2003a) Environmental correlates of walking and cycling: findings from the transportation, urban design and planning literatures. Ann Beh Med 25:80-91
[22] Saelens BE, Sallis JF, Black JB, Chen D (2003b) Neighborhood-based differences in physical activity: An environment scale evaluation. Am J Public Health 93:1552-1558
[23] Sallis JF and Frank LD, Saelens BE, Kraft MK (2004) Active transportation and physical activity: opportunities for collaboration on transportation and public health research. Transport Res A 38:249-268
[24] Sallis JF and Owen N (2002) Ecological models of health behavior. In: Glanz K, Lewis FM, Rimer BK (eds.) Health behavior and health education: theory, research and practice. Jossey-Bass, San Francisco, pp 462-484
[25] Snijder TAB, Bosker B (1993) Modeled variance in two-level models. Sociol Method Res 22: 342-363
[26] Transportation Research Board and Institute of Medicine (2005) Does the built environment influence physical activity? Examining the evidence. Special Report 282. National Academies Press, Washington DC
[27] United States Department of Health and Human Services (1996) Physical activity and health: a report of the Surgeon General. Public Health Service, Centers for Disease Control and Prevention, National Centre for Chronic Disease Prevention and Health Promotion, Atlanta GA
[27] Vandelanotte C, De Bourdeaudhuij I, Philippaerts R, Sjöström M, Sallis J (2005) Reliability and validity of a computerized and Dutch version of the International Physical Activity Questionnaire (IPAQ). J Phys Act Health 2:63-75

Objectively Assessing 'Walkability' of Local Communities: Using GIS to Identify the Relevant Environmental Attributes

Eva Leslie[a], Ester Cerin[b], Lorinne duToit[c], Neville Owen[c] and Adrian Bauman[d]

[a] School of Health and Social Development, Deakin University, Australia
[b] Institute of Human Performance, University of Hong Kong, HK, China
[c] Cancer Prevention Research Centre, University of Queensland, Australia
[d] NSW Centre for Physical Activity and Health, University of Sydney, Australia

---

Abstract: Geographic Information Systems (GIS) may be used to measure objectively, those features of the built environment that may influence walking. Public health research on environmental determinants of physical activity in adults shows that different factors can influence walking for recreation, compared to walking for transport. Most studies have used perceived (self-report) rather than objective measures of potentially relevant environmental attributes. We describe how a previously-developed index of 'walkability' was operationalized in an Australian context, using available spatial data. Attributes believed to be of relevance to walking for transport, that are measurable using GIS, are: *Dwelling density* (higher-density neighborhoods support greater retail and service variety, resulting in shorter, walkable distances between facilities; driving and parking are more difficult and time consuming). *Connectivity* (higher intersection densities provide people with a greater variety of potential routes, easier access to major roads where public transport is available and shorter times to get to destinations). *Land use mix* (the more varied the land use mix and built form, then the more conducive it is to walk to various destinations). *Net retail area* (there are more options for destinations where goods and services may be purchased and more local employment opportunities that can be reached by walking). The associations of these attributes with walking behaviors can be examined separately, or in combination. Such GIS data are very helpful in fundamental studies of the environmental determinants of behavior, and also in applied policy research for cities, regions or local communities, to address public health and environmental issues.

Keywords: GIS, community walkability, walking for transport, environment and public health.

1 Introduction

Physical inactivity is a major risk factor for overweight and obesity, diabetes, heart disease and some cancers (USDHHS, 1996). In Australia, some 30 % of adults are sedentary in their leisure time (Owen & Bauman, 1997) and over 50 % are insufficiently active to accrue health benefits (Booth et al, 1997). Environmental and policy interventions form one of the major strategic approaches to promoting participation in physical activity in industrialized countries (Sallis et al, 1998).

Researchers in both public health and the urban planning and transportation fields have highlighted the importance of using objective measures to help better understand the relationships between physical environment attributes and physical activity behaviors (Saelens et al, 2003; Sallis et al, 2004; Owen et al, 2004). Walking is the physical activity behavior that is currently the main focus of environmental and policy initiatives in public health (Owen et al, 2004). The 'walkability' of a community may be conceptualized as the extent to which characteristics of the built environment and land use may or may not be conducive to residents in the area walking for either leisure, exercise or recreation, to access services, or to travel to work. GIS provides methods that have the potential to facilitate the development of indices of walkability at the local level in cities or regional areas, not only for the purposes of research, but also to evaluate new environmental and policy initiatives (Sallis et al, 1998; Bauman et al, 2002).

Behavior-specific ecological models in population health identify a key role for particular environmental attributes as determinants of specific health-related behavioral choices (Bauman, Sallis & Owen, 2002; Sallis & Owen, 2002). In the case of adults' walking in local communities, the choice to walk may be shaped by the cueing properties of environmental attributes (for example, sidewalks, shade, accessible destinations); and, by the reinforcing consequences of walking (for example, completion of errands, use of recreational facilities) that are likely to increase the future probability of choices to walk. Studies from the community design disciplines (particularly transportation and urban planning research) have identified some strong patterns of association (Frank et al, 2003). To date, there are only a limited number of physical activity studies that have used environmental variables derived objectively from GIS data (Frank et al, 2005; Giles-Corti, 2002, 2003; Kirtland et al, 2003; Troped et al, 2001, 2003). A feature of these studies is a specificity of focus (both on the environmental attributes and the behavior) as there is an increasing body of evidence on the correlates of specific behavioral physical activities and on the role of particular environmental attributes (Humpel et al, 2002; Owen et al, 2004; Trost et al, 2002). In the case of walking, different physical attributes of local walking environments may be related to walking for particular purposes, such as walking for exercise, pleasure or transport.

Building on research findings from the transportation, urban design and planning fields, Saelens, Sallis and Frank (2003) argue that particular physical elements of local environments influence residents' choices to walk. They describe aspects of local physical environments that are considered to influence walking for transport; two main dimensions of the way land is used appear to be important: proximity (distance) and connectivity (directions of travel). *Proximity* reflects two key land-use variables: density or compactness of land use; and, land use mix (the degree of heterogeneity with which functionally different uses are co-located in space). The more compact and intermixed an urban environment is, the shorter the distances between destinations. Walking, relative to other modes of travel becomes less probable, as distances between origins and destinations increase (Frank, 2004). *Connectivity* is the directness of routes between households, stores, and workplaces. Walking is facilitated where there is a lack of barriers (freeways and other physical obstacles) and where there are a number of options for travel routes. Interconnecting streets laid out in a regular grid pattern will act to facilitate walking for transport (Saelens, Sallis & Frank, 2003; Frank, Engelke & Schmid, 2003).

The aim of this chapter is to describe the construction of an objective, GIS-based measure of community walkability (walkability index) based on key environmental characteristics related to walking for transport. This index of walkability is designed to reflect utilitarian forms of walking to work, shops and other daily destinations rather than walking for recreation and leisure. The applications, limitations and potential for further development of the walkability index are discussed.

2 Methods

2.1 Using Australian GIS data to derive a walkability index

The potential walkability dimensions of proximity and connectivity described above can be readily operationalized using GIS methods. A number of approaches have previously been used to measure walkability and the connectivity of different neighborhood designs (Aultman-Hall et al,

1997; Cervero & Kockelman, 1997; Handy, 1996; Hess, 1997; Greenwald & Boernet, 2001; Randall & Baetz, 2001). The spatial index of walkability described in this chapter is built upon the method originally developed by Frank and colleagues in the USA (Frank et al, under review) for use in the Neighborhood Quality of Life Study (NQLS; see www.nqls.org). The USA and Australian research teams collaborated on the adaptation of the walkability index for use with readily available Australian GIS databases. The index was subsequently used to generate a sampling frame for the PLACE (Physical Activity in Localities and Community Environments) study, conducted in Adelaide, the capital city of South Australia (Leslie et al, in press).

2.2. Measures

Spatial data sets used included tax valuation and cadastral (parcel) data, street centerline data, land use, zoning data, shopping centre location data integrated within a GIS to create an environmental characteristic index at the Census Collection District (CCD) level. In Australia, a CCD is the smallest unit for which population census data is made available. The following sections describe how each attribute was measured. The implied relationship to walking for transport for each of the measures calculated for the walkability index are summarized in Table 1.

Table 1. Elements of environmental characteristics and relationships to walking behavior (*From Leslie et al, in press*)

Environmental attribute	Implied Relationship with Walkability	GIS Databases to Identify the Attribute
Dwelling density	▪ High-density neighborhoods include mixed-use development - improves accessibility to variety of complementary activities and thus, increased utility ▪ Associated with increases in retail and service variety, resulting in shorter, walkable distances between complementary shops and restaurants ▪ Driving and parking more difficult and time consuming	Dwellings data
Connectivity	▪ Higher intersection densities are correlated with increased network connectivity, thus providing people with a greater variety of potential routes ▪ Higher connectivity provides easier ac-	Road centre line and intersections data

	cess to major roads where public transport is available ▪ Shorter times to get to destinations	
Land use accessibility and diversity of uses (land use attributes)	▪ People who live near multiple and diverse retail opportunities tend to make more frequent, more specialized and shorter shopping trips, many by walking ▪ People who live farther away from retail opportunities are more likely to chain together multiple shopping destinations, and to use a car ▪ The more varied the land use mix, the more varied and interesting the built-form, then the more conducive it is to walk to various destinations	Land use data and zoning data, Shopping Centers
Net area retail	▪ More options for destinations where goods and services may be purchased ▪ More local employment opportunities that can be reached by walking	Shopping Centers

2.2.1 Dwelling density

Fig. 1. Map section showing dwelling density point file classified by dwelling value with roads shown for context.

A point-based dwelling layer was created by selecting residential land use from the valuation data and summing the land area and counting the dwelling number for each CCD (See Figure 1). Net residential dwelling density was then created by dividing the dwelling count per CCD by the sum of residential land area per CCD (number of dwelling units per km^2). A standard measure was obtained by classifying into deciles.

2.2.2. Street connectivity

Intersections are identified from the street centerline data and connectivity is based upon the number of unique street connections at each intersection (or the potential for different route choices available at each intersection). Only intersections with 3 or more unique intersecting streets were included in the intersection density calculation. Density is measured based on the number of intersections per square kilometer within each CCD. As was the case with the dwelling data, intersection density is calculated by CCD (intersection count divided by CCD area) and the resultant densities classified into deciles.

2.2.3. Land use mix

The land use measure is the most complex to create and calculate and uses two data sets, land-use and zoning. Land use is the activity that is taking place on the land parcel and is the basis for levying land taxes. Zoning is the means used by local government to control the use of land and is typically grouped into broad classes such as residential, industrial, retail, recreation etc. In the land use classification it is valid to have vacant land, or land which is not being used and therefore will not attract a high land tax or council rate (See Figure 2).

Although the land is not being used, local government will have zoned the land for a specific land use that will control the activity on the land once development occurs. As vacant is a valid land use and these can be large areas that encompass many land use zones, vacant does not provide a classification that can be readily used for measuring land use mix.

Fig. 2. Map section showing the Planning South Australia land use groupings. (Notice the large area in the top left is classified as vacant under land use).

To overcome the problem of large vacant areas and the potential to skew the land use mix calculation, the underlying local government zoning is used to reclassify vacant parcels to the local government zoning (See Figure 3). In this way, the intended land use (zoning) is used in place of the vacant land use for the purpose of calculating the land use mix measure. Once the vacant land is replaced with a land use zone, the land uses are reclassified into the following five classes, residential, commercial, industrial, recreation and other.

The sum of land area by CCD is used to create an entropy score for each CCD, calculated via the following formula, where k is the category of land use; p is the proportion of the land area devoted to a specific land use; N is the number of land use categories:

$$-\frac{\sum_k (p_k \ln p_k)}{\ln N}$$

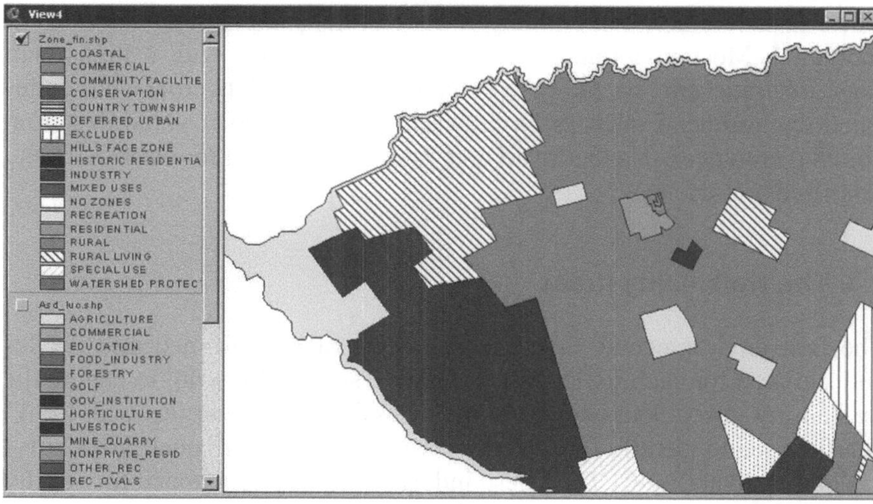

Fig. 3. Map section showing the Planning South Australia zoning (intended land use) groupings. Note the large vacant area (top left) from Fig.2 is designated as rural living under zoning.

The entropy equation results in a score of 0 to 1, with 0 representing homogeneity (all land uses are of a single type), and 1 representing heterogeneity (the developed area is evenly distributed among all land use categories). The entropy land use score for each CCD is joined to the CCD theme and classified into deciles with the 1st decile recoded to 1 representing CCDs with the least land use mix and the 10th decile recoded to 10 representing CCDs with the most mixed land uses.

2.2.4. Net area retail

The final measure in the walkability index is net retail area. A Retail Data Base is a collection of all retail activity in shopping centers with three or more shops or a single shop 250 square meters or larger. Field survey teams visit all centers and measure the gross retail area, the parcel area, retail activity and a range of other data. It is the gross retail area and the parcel area that are used in this measure as a simple ratio: NRA = GRA / P; (where GRA = gross retail area; and, P = total retail parcel area).

The premise for this measure is to calculate the amount of retail floor area in relation to the total amount of land area that serves retail use. As a result, this measure captures the degree to which retail is located near the roadway edge, as is the case in a pedestrian oriented community, or set behind a sea of parking. A ratio of 1 (retail space=total parcel space) or greater than 1 indicates retail areas where less space is devoted to cars and where distances between building entrances, transit, and other activities as shortened. Most importantly, pedestrians in locations with a high retail floor area ratio are less likely to be confronted with dangerous auto dominated environments such as large parking lots that sever sidewalks from building entrances. The net retail area ratio is calculated for each CCD and classified into deciles.

2.3. The walkability index

The four environmental characteristics were all transformed into deciles and summed for each CCD, with a possible score of 4 to 40, to denote the walkability index. This score was then classified into quartiles with the 1st quartile used to identify low walkability CCDs and the 4th quartile identifying high walkability CCDs. These indexes can then be mapped to visually identify areas at the extremes of walkability.

3 Results

Field validation was conducted to check the performance of the walkability index and determine how well the method worked as a means of sorting areas on the basis of environmental attributes which support walking behavior. The majority of selected areas' physical characteristics accorded with the classification of walkability objectively derived in GIS. Through an extensive field observation process, the study team determined that the methodology does return a face-valid means of identifying areas based upon the input characteristics of dwelling data, land use and zoning, intersection data and net retail area.

In addition, an earlier study comparing residents' perceptions of walkability attributes in objectively different neighborhoods at the extremes of walkability, found that residents of high and low-walkable neighborhoods could reliably perceive the objective differences between them (Leslie et al, 2005). This earlier study used only three components for measuring walkability (net retail area was not included). Figure 4 shows the mean subscale scores for residents in high and low walkable neighborhoods. Comparison of scores shows that residents in the high-walkable neighborhoods provided ratings indicative of higher residential density, land use mix-diversity, land use mix access, street connectivity and infrastructure for walking (all $p<.001$) than did residents of low-walkable neighborhoods.

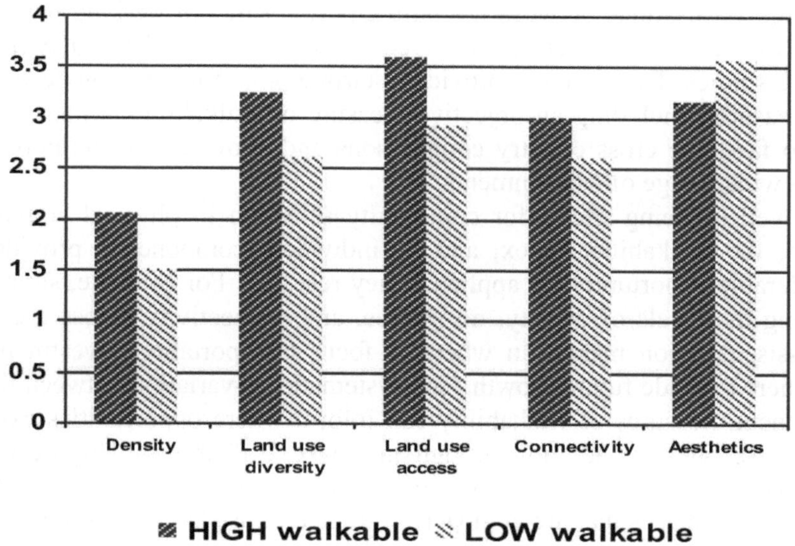

Fig. 4. Comparison of mean scores for perceived environmental characteristics for residents in high- and low-walkable neighborhoods (all statistically significantly different, $p<.001$). *Adapted from Leslie et al, 2005.*

4 Discussion

The four environmental characteristics utilized in the method described above are a starting point to a more detailed and informed measure of walkability. There are plausibly many other factors linked with walking behavior that warrant further consideration. However, it should be noted that while there may be a range of factors that are worthy of consideration, data is specific to particular areas and it may not be possible to include a large range of the same factors across different areas.

Applications

The walkability index and its component measures, provides a useful tool for the selection of communities for household recruitment, as it can maximize the variability in the built environment and result in an improved ability to detect differences in physical activity levels that likely occur in objectively different environments. Selecting communities based on extremes of walkability is currently being promoted by the International Physical Activity and the Environment Network (www.ipenproject.org) to study the associations between the built environment and walking behaviors. Two key studies that are involved in this work are the NQLS and PLACE studies. These studies provide a starting point for common designs and measures, including an objective measure of walkability, which will help to facilitate cross country comparisons and allow data to be pooled from a wider range of environments.

As well as being useful for community selection in physical activity studies, the walkability index, and its individual components, provides considerable opportunity for applied policy research. For example, simply mapping the levels of density, mixed use, and connectivity across a city can assist decision makers in where to focus transportation investments and where to guide future growth. The systematic co-variation between the component measures of walkability can inform where opportunities exist to increase physical activity through improved walkability. This occurs where considerable disparities exist between the measures of walkability. For example, places that are compact and mixed in use but offer little ability to traverse between destinations in a direct manner can be detected and targeted for improvements in street connectivity through various approaches to transportation investment including connecting cul-de-sacs, and the completion of sidewalk systems. Places that have high levels of mixed use and good connectivity may be targeted for more residential development and increased density. This would lead to more pedestrian friendly and transport efficient urban design and would increase opportunities for 'active living' in communities. The identification of targeted areas for investments that have the potential to offset auto use and relieve traffic congestion and air pollution would be seen as cost effective and would help address contemporary public health and environmental issues (Frank et al, 2004; 2005).

Limitations

A limitation of the method used here is that there are likely a number of other related environmental attributes plausibly related to walking for transport that were not included. These include (but are not limited to) the presence, condition and continuity of footpaths (Handy et al, 2002; Cervero & Kockelman, 1997), accessibility or distances to facilities (Pikora et al, 2002; Giles-Corti & Donovan, 2002;2003), transit accessibility (Kitamura et al, 1997), factors related to natural features such as topography and physical barriers (Rodriguez & Joo, 2004), and other aspects of urban design such as building design and orientation, street lighting, planted strips. A more detailed and informed measure of a walkability index could include some or all of these characteristics. Future analyses could examine a wider range of GIS variables for their relation to physical activity, with the eventual aim of creating a more comprehensive walkability index.

As with any use of GIS data there are also issues of data quality and integrity. For example, boundaries and attributes may no longer match if the data has been collected at various timepoints. Some data may be incomplete or may include variation as a result of being constructed at various stages and accuracy for some attributes could be ambiguous or may lack sufficient detail. There will naturally be a need to make assumptions and decisions along the way in working with any 'raw' data and this will vary according to who is working with it.

Further development of the walkability index

A further development in the application of the walkability index in public health, transport and planning research is to use these characteristics at the individual residence level. Many of the elements involved in the derivation of the index of walkability are applicable at an individual residence level and it would be desirable to calculate the index for each individual residence. This would require identifying the relevant boundaries around each residential location, within which the measures of relevant attributes can be derived and an individual walkability score calculated. This would allow a surface of walkability to be derived which could then be aggregated to any desired spatial level to protect confidentiality and for further analysis. For example, new measures that allow for assessment of walking within (10-15 minute walk from home) and outside local neighborhoods have recently been developed (Giles-Corti et al, 2006). These can be used against GIS methods that increase the specificity of a neighborhood walkability measure by using smaller spatial units based on a resident's location. Methods for increasing such specificity are currently being developed (Learnihan et al, 2006). It is anticipated that refined measures that include additional elements and that build upon the actual residential location may provide a more robust measure with a greater range of inputs, be spatially more flexible and can be aggregated to a multitude of geographic scales at the observation, community, sub-area, and regional scale.

5 Conclusion

There are a wide array of potential applications of GIS methods for guiding environmental and policy initiatives to promote walking and to increase overall physical activity levels. Increased computing capabilities, in concert with the availability of GIS based land use and transportation data provide considerable opportunity to develop objective measures of the built environment that form independent predictors of human activity patterns.

Acknowledgements

We thank Neil Coffee and the National Centre for Social Applications of GIS (GISCA) at the University of Adelaide for technical support. The National Health and Medical Research Council (NHMRC) Program Grant #301200 supported the PLACE (Physical Activity in Localities and Community Environments) study. Dr Leslie is supported by an NHMRC Public Health Fellowship # 301261.

References

[1] Aultman-Hall L, Roorda M, Baetz BW (1997) Using GIS for evaluation of neighborhood pedestrian accessibility. *Journal of Urban Planning and Development* 123(1):10-17

[2] Bauman A, Sallis JF, Owen N (2002) Environmental and policy measurement in physical activity research. In Welk G & Dale D (Eds.), *Physical activity assessments for health-related research.* Champaign, Illinois: Human Kinetics, pp. 241-251

[3] Booth ML, Bauman A, Owen N, Gore CJ (1997) Physical activity preferences, preferred sources of assistance and perceived barriers to increased activity among physically inactive Australians. *Preventive Medicine,* 26:131-7

[4] Cervero R and Kockelman K (1997) Travel demand and the 3Ds: Density, diversity, and design. *Transportation Research Part D,* 2(3):199-219

[5] Frank LD (2004) Economic determinants of urban form. Resulting trade-offs between active and sedentary forms of travel. *American Journal of Preventive Medicine* 27(3S): 146-153

[6] Frank L, Andresen MA, Schmid TL (2004) Obesity relationships with community design, physical activity, and time spent in cars. *American Journal of Preventive Medicine* 27(2):87-96

[7] Frank LD, Engelke PO, Schmid TL (2003) *Health and community design. The impact of the built environment on physical activity*. Washington: Island Press

[8] Frank LD, Sallis JF, Saelens BE, Leary LE, Cain K, Conway T (*manuscript under review*). Assessing the relation between urban form and physical activity: design and methods of the Neighborhood Quality of Life Study

[8] Frank LD, Schmid TL, Sallis JF, Chapman J, Saelens BE (2005) Linking objectively measured physical activity with objectively measured urban form. *American Journal of Preventive Medicine* 28(2S2):117-125

[9] Giles-Corti B and Donovan RJ (2002) The relative influence of individual, social and physical determinants of physical activity. *Social Science and Medicine* 54(12):1793-1812

[10] Giles-Corti B and Donovan RJ (2003) Relative influences of individual, social environmental and physical environmental correlates of walking. *American Journal of Public Health* 93(3):1583-1589

[11] Giles-Corti B, Timperio A, Cutt H, Pikora TJ, Bull FCL, Knuiman M, et al. (2006) Development of a reliable measure of walking within and outside the local neighborhood: RESIDE's Neighborhood Physical Activity Questionnaire. *Preventive Medicine* 42:455-459

[12] Greenwald MJ and Boarnet MG (2001) The built environment as a determinant of walking behavior: Analyzing non-work pedestrian travel in Portland, Oregon. *Transportation Research Record* 1780:33-43

[13] Handy SL, Boarnet MG, Ewing R, Killingsworth RE (2002) How the built environment affects physical activity. Views from urban planning. *American Journal of Preventive Medicine* 23(2S):64-73

[14] Hess PM (1997) Measures of connectivity. *Places* 11:58-65

[15] Humpel N, Owen N, Leslie E (2002) Environmental factors associated with adults' participation in physical activity: A review. *American Journal of Preventive Medicine* 22(3):188-99

[16] Kitamura R, Mokhtarain PL, Laidet L (1997) A micro-analysis of land use and travel in five neighborhoods in the San Francisco Bay area. *Transportation* 24:125-128

[17] Kirtland KA, Porter DE, Addy CL, Neet MJ, Williams JE, Sharpe PA, et al. (2003) Environmental measures of physical activity supports: Perception versus reality. *American Journal of Preventive Medicine* 24(4):323-331

[18] Learnihan V, Van Niel K, Giles-Corti B (2006) Walking and the built environment. The influence of data scale in walkability analyses: results from *RESIDE*. Abstract in Fifth Conference of the International Society for Behavioral Nutrition and Physical Activity, ISBNPA, Boston, USA, p.240

[19] Leslie E, Saelens B, Frank L, Owen N, Bauman A, Coffee N, et al. (2005). Residents' perceptions of walkability attributes in objectively different neighborhoods: a pilot study. *Health and Place* 11:227-236

[20] Leslie E, Coffee N, Frank L, Owen N, Bauman A, Hugo G (in press, available on line). Walkability of local communities: using Geographical Information Systems to objectively assess relevant environmental attributes. *Health and Place*

[21] Owen N and Bauman A (1997) The descriptive epidemiology of a sedentary lifestyle in adult Australians. *International Journal of Epidemiology* 21(2):305-10

[22] Owen N, Humpel N, Leslie E, Bauman A, Sallis JF. (2004) Understanding environmental influences on walking: review and research agenda. *American Journal of Preventive Medicine* 27(1):67-76

[23] Pikora TJ, Bull FCL, Jamrozik K, Knuiman M, Giles-Corti B, Donovan RJ (2002) Developing a reliable audit instrument to measure the physical environment for physical activity. *American Journal of Preventive Medicine* 23(3):187-194

[24] Randall TA and Baetz BW (2001) Evaluating pedestrian connectivity for suburban sustainability. *Journal of Urban Planning and Development* 127:1-15

[25] Rodriguez DA and Joo J (2004) The relationship between non-motorized mode choice and the local physical environment. *Transportation Research Part D,* 9:151-173

[26] Sallis J, Bauman A, Pratt M (1998) Environmental and policy interventions to promote physical activity. *American Journal of Preventive Medicine* 15(4): 379-97

[27] Saelens BE, Sallis JF, Frank LD (2003) Environmental correlates of walking and cycling: Findings from the transportation, urban design, and planning literatures. *Annals of Behavioral Medicine* 25(2):80-91

[28] Sallis JF, Frank LD, Saelens BE, Kraft MK (2004) Active transportation and physical activity: opportunities for collaboration on transportation and public health. *Transportation Research Part A,* 38:249-268

[29] Sallis JF and Owen N (2002) Ecological models of health behavior. In: Glanz K, Lewis FM, Rimer BK. (eds.) *Health Behavior and Health Education: Theory, Research, and Practice.* 3nd ed. San Francisco: Jossey-Bass, pp. 462-484

[30] Trost SG, Owen N, Bauman AE, Sallis JF, Brown W (2002) Correlates of adults participation in physical activity: review and update. *Medicine and Science in Sports and Exercise* 34(12):1996-2001

[31] Troped PJ, Saunders RP, Pate RR, Reininger B, Addy CL (2003) Correlates of recreational and transportation physical activity among adults in a New England community. *Preventive Medicine* 37:304-310

[32] United States Department of Health and Human Services (1996) *Physical Activity and Health: A Report of the Surgeon General.* Atlanta, GA: Public Health Service, Centers for Disease Control and Prevention, National Centre for Chronic Disease Prevention and Health Promotion

Developing Habitat-suitability Maps of Invasive Ragweed (*Ambrosia artemisiifolia.L*) in China Using GIS and Statistical Methods

Hao Chen[a,b], Lijun Chen[b], and Thomas P. Albright[c]

[a] School of Remote Sensing Information Engineering, Wuhan university, Wuhan, China,
[b] National Geomatics Centre of China, Beijing, China,
[c] Department of Zoology, University of Wisconsin-Madison, WI, USA

Abstract: Invasive alien species pose a large and growing threat to the economy, public health, and ecological integrity in China. Ragweed (Ambrosia artemisiifolia.L), native to North America, was first documented in China in 1935. As the primary pathogen of pollinosis, ragweed has become a serious menace to public health of China's population. Explaining and predicting the spatial distribution of ragweed is of great importance to prevention and early warning efforts. We studied the potential distribution of ragweed, the environmental factors that influence its distributions, as well as the ability to compute habitat-suitability maps using GIS and statistical tools. We have developed models using logistic regression based on herbarium specimen locations and a suite of GIS layers including climatic, topographic, and land cover information. Our logistic regression model was based on Akaike's Information Criterion (AIC) from a suite of ecologically reasonable predictor variables. Based on the results, we provided a new method to compartmentalize the habitat-suitability in the native range. Finally, we used the model and the compartmentalized criterion developed in the native range to "project" a potential distribution onto the exotic ranges to build habitat-suitability maps.

Keywords: invasive species, habitat-suitability maps, Akaike's Information Criterion (AIC), logistic regression, quantile, GIS

1 Introduction

Invasive alien species pose a growing threat to the economy, public health, and ecological integrity of nations worldwide. The invasions on local ecosystem functions and processes of foreign species are increasingly recognized in China, especially when the Chinese population, trade, and travel continue to grow (Guo 2002). In order to effectively control these invasive species, a better understanding of the nature of the species and species–environment relationship is required.

One approach to the quantification of such species–environment relationships involves the use of habitat-suitability or niche-based models whereby environmental conditions suitable for sustaining the population of a species are identified (Peterson 2003). This approach combines herbarium specimen location data with a suite of GIS layers (e.g. climatic, topographic, and land cover) to create ecological models of the species' requirements. Coupled with these models, GIS can project the ecological model onto a geographic space and map the habitat-suitability maps in native and exotic ranges (Peterson 2003).

Ordinary multiple regression and generalized linear models (GLM) are very popular and are often used for modeling species niche. The traditional approach using GLM has been used for hypothesis testing. An alternative strategy for model selection and inference is based on information theoretic approaches and uses metrics such as Akaike's information criteria (AIC). This approach offers several advantages for modeling species distributions, including the abilities to compare non-nested models, determine relative variable importance, and perform multi-model inference. In addition, many researchers find its simplicity and lack of a need to designate an arbitrary or conventional p-value appealing (Greaves et al. 2006).

Traditional logistic regression requires good quality presence/absence data in order to generate a probability surface of habitat suitability (Manel et al. 1999, Guisan and Zimmerman 2000). In practice, accurate data on the absence of a species is difficult to obtain as most museum databases record locations where species were collected but not where species were nonexistent, Absence data may be questioned even in the field because it can be determined quickly if a given plant is in a plot, but determining absence requires a much more thorough search than is practical for a broad scale study. One solution is to generate "pseudo-absence" data by selecting points at random from the geographical space. This approach has the risk of creating absence sites where a species is actually but unknown to be present. A better method is required for habitat suitability modeling using the GLM approach given the unavailability of true absence data.

In this study, we employed the information-theoretic and stepwise selection approaches to generate the logistic models for predicting the occurrence of ragweed. The aim was to assess the modeling choice as well as to produce a model that would predict the suitability of ragweed in both native and invaded ranges. Much of the available data on species occurrence consists only of presence data sets. In order to decrease this sampling bias, we proposed a new approach specifically to compartmentalize the habitat-suitability using logit value threshold and the quantile statistics.

2 Methods

2.1 Study Species

Ragweed, originally found in North America, is an annual herbaceous plant. It is a well-documented cause of seasonal allergic rhinitis and seasonal asthma for countries in the Northern Hemisphere such as the United States and Canada (Creticos et al. 1996, Boulet et al. 1997). It appeared that Ragweed accidentally intruded china and was discovered in northeast China for the first time in 1935. It then rapidly spread to northern and central China, covering 15 provinces, but its ability to continue its invasion into other regions of China is unknown. By 2004 and in Jilin province alone, there were over 150,000 people suffering from the pollinosis related to ragweed with more than 530,000 hectares of farmland threatened by resource competition from ragweed (Cui et al. 2004). The aim of this chapter was to predict the potential future distribution of this species based on its habitat suitability.

2.2 Species Data

Habitat-suitability models were based on 243 unique georeferenced herbarium specimen records derived from the native range (USA and Canada) of ragweed accumulated from museum collections and databases (see Acknowledgments) An additional 83 points from the species' invaded range (China) were gathered from the herbarium museum of the Kunming Institute of Botany and the herbarium museum of the Institute of Botany of the Chinese Academy of Sciences. The occurrences of ragweed are shown in Figure 1. Because we have no absence data for ragweed, we generated "pseudo-absence" data by selecting points at random from the geographical space in the native range. All occurrence points were georeferenced to a precision finer than the 0.167 degree of latitude and longitude grain size used for the modeling.

Fig. 1. Distribution of herbarium specimen records for ragweed (•) in native (North America) and invaded ranges (China)

2.3 Selection of Habitat Variables

Many potential predictors were either different measures of the same ecological characteristic or were calculated using the same data, thus inheriting problems of collinearity (Legendre and Legendre 1998). For this reason, six biologically relevant environmental factors were considered for inclusion in the analysis (Table 1). They included elevation (Elev), mean annual precipitation (Precip), minimum temperature of the coldest month (Mintmp), percent of maximum solar radiation (Sun), the proportion of areas devoted to either agricultural or urban land uses (Agurb), and percent forest cover (Fordens). Although these data were collected from sources of variable scales, they were all converted to the same resolution level of 0.167°×0.167° (about 15×15 km2 for the middle latitudes of China).

Altitude was selected as a variable because of its influence in the climatic conditions of the ragweed that result in changes to the vegetation and the timing and length of fruit bearing and flowering seasons. Precipitation has the potential impact on soil moisture and the water balance of ragweed. Minimum temperature and solar radiation are both essential factors to the growth of ragweed. The interaction terms including (Sunshine * Minimum_temperature) and (Precipitation * Minimum_temperature) may be important for the plant in periods of short daylight. The four coverages were all from CRU CL 2.0 (New et al. 2002) and Worldclim 1.4 (http://biogeo.berkeley edu/worldclim/worldclim.htm). Human disturbance is very important for the expansion and persistence of many invasive plants, including ragweed, and the percentage of agriculture and urban landuse (derived from the USGS Global Landcover Characterization dataset) (Loveland *et al.* 2000) was thought to be a reasonable proxy for this. Finally, density of forest cover came from the UN/FAO Forest Resources Assessment 2000 dataset (Waller and Zhu 1999).

Table 1. Prior knowledge of predictor variables and their abbreviations used in the model

Abbreviation	Predictor Variables	Description
Elev	Elevation	Elevation may influence associated climatic conditions.
Precip	Precipitation	Precipitation has the potential to influence soil moisture and the water balance of ragweed.
Mintmp	Min temperature of coldest month	Ragweed is very sensitive to low temperatures.
Sun	Sunshine	Percent of maximum possible solar radiation is very important for the plant under short days.
Agurb	Density of agriculture and urban	Human settlement may influence the spread of ragweed.
Fordens	Density of forest	Treeless areas may not support population. On the other hand, very high forest cover may limit opportunities.

2.3 Statistical Analyses

Binary Logistic Regression

We used the GLM to predict the potential distribution of ragweed based on presence points and "pseudo-absence" data in their native range. As the response variable of this analysis was binary, the appropriate form of GLM was binomial logistic regression (Guisan and Zimmermann 2000). The probability of the species occurrence is calculated as follows:

$$probability(even) = \frac{e^{B_0+B_1X_1+B_2X_2...B_KX_K}}{1+e^{B_0+B_1X_1+B_2X_2...B_KX_K}}, \qquad \text{Eqn. 2.1}$$

where B_0 to B_k are coefficients, X_1 to X_k are the independent variables. Logistic regression analysis was in the R statistical package. Initially three variables (Precip, Mintmp, and Sun) and their second order terms were included in a core model. We used a stepwise approach to add the other variables to the basic model to build more complicated models using logistic regression. The additional variables included Elev, Agurb, Fordens and their second order terms and two interaction terms: Sun* Mintmp and Precip * Mintmp.

Akaike's Information Criterion (AIC)

The selection of the logistic regression model was based on Akaike's Information Criterion (AIC). AIC represents the relationship between the maximum likelihood and Kullback-Leibler information. It is based on the principle of parsimony (by including a penalty for the number of parameters) and attempts to select a good approximating model (Anderson et al. 2000, Burnham and Anderson 2002). AIC is defined as:

$$\mathrm{AIC} = -2\log_e\left(\mathrm{L}\left(\hat{\theta} \mid \mathrm{data}\right)\right) + 2\mathrm{K}, \qquad \text{Eqn. 2.2}$$

where $-2\log_e\left(\mathrm{L}\left(\hat{\theta} \mid \mathrm{data}\right)\right)$ is the value of the maximized-log-likelihood over the unknown parameters ($\hat{\theta}$), given the data and the model, and K is the number of estimable parameters included in the model (number of variables + the intercept).

The models can then be ranked from best to worst using the delta AIC (Δi), which is a measure of each model relative to the best model, and is calculated as:

$$\text{Delta AIC} = \Delta_i = \mathrm{AIC}_i - \min \mathrm{AIC}, \qquad \text{Eqn.2.3}$$

where AIC_i is the AIC value for model i , and minAIC the AIC value of the best model. We ran each of the models and computed the AIC and Δi in the R statistical package. Akaike weights (wi) provide another measure of the strength of evidence for each model, and represent the probabilities that model i is the Kullback–Leibler best model in the whole set of *R* candidate models:

$$\mathrm{w}_i = \frac{\exp\left(-\Delta_I/2\right)}{\sum_{r=i}^{R} \exp\left(-\Delta_I/2\right)}. \qquad \text{Eqn. 2.4}$$

A weighted average was calculated for the coefficient of each parameter $\overline{\hat{\theta}}$ for the average model using the equation below, $\hat{\theta}_i$ is the coefficient for model i.

$$\text{model-averaged estimate} = \overline{\hat{\theta}} = \sum_{i=1}^{R} w_i \hat{\theta}_i. \qquad \text{Eqn. 2.5}$$

The assessment of relative importance of the predictor variables can be made by summing the Akaike weights for all models containing a given predictor variable rather than a single selected model (Burnham and Anderson 2001).

2.4 Habitat-suitability Compartmentalization

Calculation of the Logit Value

Having obtained the weighted average model and their regression coefficients (see Table 3), the resulting regression equations were applied to the native samples including the presence points and "pseudo-absence" points to produce the output of the logit value. The final function takes the form of:

$$\begin{aligned}\log it = {} & \beta_0 + \beta_1 * \text{precip} + \beta_2 * \text{precip}^2 + \beta_3 * \text{mintmp} + \beta_4 * \text{mintmp}^2 + \beta_5 * \text{sun} \\ & + \beta_6 * \text{sun}^2 + \beta_7 * \text{elev} + \beta_8 * \text{elev}^2 + \beta_9 * \text{fordens} + \beta_{10} * \text{fordens}^2 + \beta_{11} * \text{agurb} \\ & + \beta_{12} * \text{agurb}^2 + \beta_{13} * (\text{sun} * \text{mintmp}) + \beta_{14} * (\text{precip} * \text{mintmp})\end{aligned} \quad \text{Eqn.2.6}$$

Re-class for the Samples

Because of the lack of true absence data, it was not be possible to calculate meaningful probabilities from the logit values. Instead, we developed a method, inspired by resource selection functions (Manly et al. 2002), to map relative suitability values based on logit thresholds and quantiles.

Firstly, we defined a suitability threshold based on the minimum logit value of reference presence points in the native range. As an alternative, we also determined this threshold from the invaded range data. The threshold allowed us to place every pixel broadly into either suitable or unsuitable categories depending on whether its modeled logit value exceeded or was less than the threshold. Finally, we calculated quartiles for each of these two groups using the quantile function (Hyndman and Fan 1996) in the R statistical package. The resulting eight classes ranked from the most unsuitable to the most suitable are as below:

Unsuitable	< minimum logit
1 most unsuitable	< first quartile of unsuitable samples
2	first quartile of unsuitable samples ~ second quartile of unsuitable samples.
3	second quartile of unsuitable samples ~ third quartile of unsuitable samples
4 slightly unsuitable	third quartile of unsuitable samples ~ minimum logit
Suitable	> minimum logit
5 slightly suitable	minimum logit ~ first quartile of suitable samples
6	first quartile of suitable samples ~ second quartile of suitable samples
7	second quartile of suitable samples ~ third quartile of suitable samples
8 most suitable	> third quartile of suitable samples

Compute the Habitat-suitability Map

Logit values were then calculated for every pixel in ERDAS 8.6 for both native (North America) and invaded (China) ranges, according to the Equation 2.6 and Table 3. The resulting maps of logit values were then categorized into the eight suitability classes.

Because of the geographic and floristic similarities between China and North America, we used the habitat-suitability models developed from native ranges to "project" onto the exotic ranges to predict the ragweed's potential distribution in China. But in view of variation between native and invasive ranges, we tried to use the minimal logit value of herbarium specimen locations in invasive (China), instead of the native (North America), ranges as the critical value to compute habitat-suitability in order to reflect the current distribution of ragweed in China.

3 Results

3.1 Logistic Regression

In this chapter, we used a conservative strategy for selecting models using information theoretic approach. Initially based on the prior knowledge, this was carried out using three variables (precip, mintmp, and sun) and their square terms as the core model. We then added the other variables to the core model using stepwise approach to build forty-nine more complicated models.

Table 2. Results of information-theoretic statistics for the best nine models and the core model in rank order for ragweed

Model Description[a]	K	-2LL	AIC	Δ_i	w_i
precip + precip^2 + mintmp + mintm^2 + sun + sun^2 + elev + agurb + agurb^2 + fordens + fordens^2 + sun*mintmp	13	912.42	938.42	0.00	0.2460
precip + precip^2 + mintmp + mintm^2 + sun + sun^2 + elev + agurb + agurb^2 + fordens + fordens^2 + precip* mintmp	13	910.98	938.98	0.56	0.1859
precip + precip^2 + mintmp + mintm^2 + sun + sun^2 + elev + fordens + for-dens^2+ precip*mintmp + sun* mintmp	12	916.17	940.17	1.75	0.1026
precip + precip^2 + mintmp + mintm^2 + sun + sun^2 + elev + elev^2 +agurb + agurb^2 + fordens + fordens^2 + sun* mintmp	14	912.42	940.42	2.00	0.0905
precip + precip^2 + mintmp + mintm^2 + sun + sun^2 + elev + fordens + fodens^2 + sun*mintmp	11	918.64	940.64	2.22	0.0811
precip + precip^2 + mintmp + mintm^2 + sun + sun^2 + elev + elev^2+ fordens + fordens^2+ agurb + agurb^2 + sun* mintmp + precip*mintmp	15	910.94	940.94	2.52	0.0698
precip + precip^2 + mintmp + mintm^2 + sun + sun^2 + elev + fordens + fodens^2 + agurb + sun*mintmp	12	917.51	941.51	3.09	0.0524
precip + precip^2 + mintmp + mintm^2 + sun + sun^2 + elev + elev^2+ fordens + fordens^2+ sun*mintmp + precip* mintmp	13	916.00	942.00	3.58	0.0411
precip + precip^2 + mintmp + mintm^2 + sun + sun^2 + elev + elev^2+ fordens + fordens^2+ sun*mintmp	12	918.55	942.55	4.13	0.0312
precip + precip^2 + mintmp + mintm^2 + sun + sun^2 (core)	7	975.16	989.16	50.7	0.000

a : Models are referenced by explanatory variables in their abbreviated forms (see Table 1).

Table 2 shows the number of estimated parameters in an approximating model (K), the maximized log-likelihood (-2LL), the AIC values, differences between the model with the lowest AIC value and each candidate model (Δi), Akaike weights (wi) for the best nine models and the core model. The model with the lowest AIC included the core variables, elevation, agurb, fordens, and sun*mintmp as variables but only has an Akaike weight of 0.2460, suggesting that this model might not be the best model. The top four models had Δ_i <2 and so were considered the most parsimonious (Burnham and Anderson 2002). The next four models had Δ_i<4, and the ninth model had Δ_i of just over 4, and they were considered useful for explaining variances in dependent variables. The sum of the weights for these nine models was 0.906, and thus considered as having a high confidence relative to other models considered. Besides the three variables in the core model, both elev and fordens were included in all of the top 9 models, suggesting that they were very important for improving the predictive power of the models (Greaves et al. 2006).

3.2 Model Averaging

A weighted average of the estimates was computed based on the Akaike weights of the above nine models, instead of relying solely on the first rank model. The weights for each of the variables and the coefficients for the averaged model are given in Table 3.

Table 3. The coefficients for AIC >90% certainty model average and the weights for each of the variables

Model variable	Coefficient		Importance Weight[c]
Precip	-0.00645[a]	-1.34339[b]	1.0000
Mintmp	0.13393[a]	-0.00002[b]	1.0000
Sun	0.01544[a]	4.84667[b]	1.0000
Elev	-0.00158[a]	-0.00158[b]	0.9996
Fordens	0.03388[a]	-0.00032[b]	0.9806
Agurb	0.00145[a]	-1.28464[b]	0.7533
Sun*Mintmp	-5.91351		0.9501
Precip*Mintmp	3.04714		0.4441
Intercept	-219.68		•

a: the coefficient of the first order term; b: the coefficient of the second order term; c: including both the first and second order terms; •: no calculation.

The weights for the three variables in the core model are one, because they were included in each of the nine best models. Elev and fordens both had the very high predicting weight and occurred in all high ranked models, reflecting that the two variables were very important predictors. Agurb and sun*mintmp was supported as an important variable too, whilst precip*mintmp were much less important.

3.3 Habitat-suitability Map

Having obtained the weighted average model, the resulting regression equations were applied to local data set of the native range which included the presence and "pseudo-absence" points to produce the output of the logit value. The minimal logit value and the correlative quantile thresholds calculated by the frequency statistical approach formed the habitat-suitability compartmentalization criterion for ragweed (see Table 4).

Table 4. The habitat-suitability compartmentalization criterion and the correlative threshold

Suitability Classes	Compartmentalization criterion	Coefficients[c]	Coefficients[d]
Unsuitable	< minimal logit	< -2.09197	< -0.93756
1 most unsuitable	< 1th quartile[a]	< -3.47050	< -3.17542
2	1th quartile[a] ~ 2th quartile[a]	-3.47050 ~ -2.96187	-3.17542 ~ -2.46708
3	2th quartile[a] ~ 3th quartile[a]	-2.96187 ~ -2.53550	-2.46708 ~-1.80273
4 slightly unsuitable	3th quartile[a] ~ minimal logit	-2.53550 ~ -2.09197	-1.80273 ~ -0.93756
Suitable	> minimal logit	> -2.09197	> -0.93756
5 slightly suitable	minimal logit ~ 1th quartile[b]	-2.09197 ~ -1.40881	-0.93756 ~-0.70046
6	1th quartile[b] ~ 2th quartile[b]	-1.40881 ~ -0.78746	-0.70046 ~ -0.50788
7	2th quartile[b] ~ 3th quartile[b]	-0.78746 ~ -0.42575	-0.50788 ~ -0.27703
8 most suitable	> 3th quartile[b]	> -0.42575	> -0.27703

a: the native samples (including presence and pseudo-absence) in which the logit value < the minimal logit value; b: the native samples (including present and pseudo-absence) in which the logit value > the minimal logit value; c: based on the minimal logit of herbarium specimen locations in native range (N.A.); d: based on the minimum logit of herbarium specimen locations in invaded range (China).

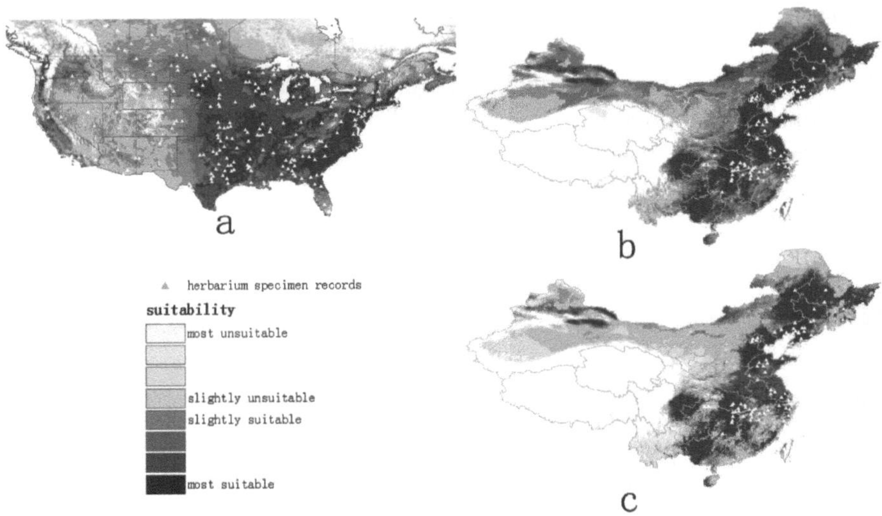

Fig. 2. Habitat-suitability maps for ragweed in both native and invaded ranges. Darker shading indicates greater confidence in the prediction of presence. The map shows logit thresholds derived from the native herbarium specimen records (*a* and *b*) and invasive herbarium specimen records(*c*).

Combining with the habitat-suitability compartmentalization criterion, the weighted average models were implemented into the GIS layers to fractionize the native range into eight classes ranked from the most suitable to the most unsuitable and generated the habitat-suitability map for ragweed in North America (Fig. 2a). We next "projected" the model developed from the native ranges onto the exotic ranges and generated the ragweed's habitat-suitability map in China (Fig. 2b). The habitat-suitability map shows that Ragweed has very strong potential to expand in China. However, the distribution appears to have been over-predicted. This problem might arise from the variation in habitat availability across the geographic ranges (Osborne and Sua´rez-Seoane 2002). We then used the minimal logit value of herbarium specimen locations in the invaded range (China), instead of that derived from the native range (U.S and Canada) to compute another habitat-suitability map for China (Fig. 2c). This map was more conservative and consistent with known records of the current ragweed distribution in China. Outside of the known current ranges of ragweed, the Sichuan Basin, Xinjiang, and areas to the southwest China were identified as the most suitable regions for ragweed. The results suggested that if given the opportunity through the spread of propagules, ragweed may invade these areas in the future.

4 Discussions

4.1 The Potential of Ragweed Spreading into New Regions in China

Both the suitability map generated by the North America-derived thresholds and the map using the more conservative China-derived thresholds indicated a strong potential of ragweed to expand into regions of China not known of its occurrences including Sichuan, Guizhou, Guangxi, Hainan, and Xinjiang.

4.2 Information-theoretic Approach

When conducting statistical analyses, we often strive to estimate the effects of a given variable on a response variable and the associated confidence level. Classical techniques such as significance tests of null hypotheses are well suited for manipulative experiments. But in recent years, the validity of this broad scale environmental method has been questioned by a number of researchers (Anderson et al. 2001, 2000; Anderson and Burnham 2002; Burnham and Anderson 2002). Null hypotheses are often biologically unlikely to occur; there being a big difference between statistical significance and biological importance (Yoccoz 1991). Furthermore, hypothesis testing does not perform particularly well in model selection, (e.g., variables selected by forward, backward, or stepwise approaches).

Information-theoretic approach built on the idea that there may be no single true model. Indeed, models only approximate reality (Kullback and Leibler 1951). We try at best to find which model would best approximate reality – that is to say minimize the loss of information as measured by Kullback-Leibler information. Akaike's information criteria (AIC) can estimate the relationship between the maximum likelihood, which is an estimation method used in many statistical analyses, and the Kullback-Leibler information. In this chapter we built forty-nine models using logistic regression and used AIC to rank this set of models on how closely they approximate reality. The Akaike weights for the top nine models (Table 3) are all relatively low (the top one model only has an Akaike weight of 0.2460) which indicated that there was too much model selection uncertainty for this to be considered the best model (Greaves et al. 2006). One of the advantages of information-theoretic approach is that it allows model averaging instead of solely dependent on the model itself. The sum of the weights of the top nine models was 0.906, which suggests a high confidence that these models together are better at explaining the observed data than the remaining candidate models. The average model calculated based on the Akaike weights was shown in Table 3. Another advantage of the Information-theoretic Approach is that it can measure the importance of individual variables based on the multiple models instead of only one chosen model, which alone may not accurately indicate variable importance.

4.3 Approach of Habitat-suitability Compartmentalization without True Absence Data

Sampling for presence/absence information is a crucial step for niche modeling. The sample must be unbiased to be representative of the whole population. In many cases, including ours, accurate data on absences is difficult to obtain, and so "pseudo-absence" data generated by selecting sample locations at random from the geographical space are used instead as true absence data. There are two types of results from logistic regression: one is the logit value that is the output of the linear equation, and the other is the probability value which reflects the occurrence probability for the species .The relationship between them is expressed below:

$$\beta_0 + \beta_1 \text{var}\,iable_1 + \beta_2 \text{var}\,iable_2 + \cdots + \beta_i \text{var}\,iable_i = \log it, \quad (4.1)$$

$$probability_{(y=1)} = \frac{\exp(\log it)}{1 + \exp(\log it)} \quad (4.2)$$

For binomial logistic regression the absence probability for the species is as below:

$$probability_{(y=0)} = 1 - probability_{(y=1)} \quad (4.3)$$

For lack of true absence data, we did not transform the logit value back to regular space scaled from 0-1 representing probability of a pixel containing the species but regard the logit value as the degree of the suitability for the species. More explicitly, we set the minimal logit value of exact present samples as the threshold to divide the whole samples into two sets of new sample points: suitable samples and unsuitable samples, then respectively calculated the quantiles for these two sets of points based on the logit value to generate a new criterion for compartmentalizing the habitat-suitability using the result of the logistic models (Table 4). In this case, the approach yielded easily interpretable and highly plausible results.

5 Conclusion

The distribution of ragweed in China is strongly influenced by precipitation (negatively), sunshine (positively) and minimum annual temperature, elevation, percent forest cover, and, to a lesser extent, the fraction of the landscape intensively used by humans. The relationship between ragweed and these factors in both its native North America and invaded range in China, as manifested in the mapped suitability strongly suggests the ability of ragweed to continue expansion in China into new areas such as Sichuan, Guizhou, Guangxi, Hainan, and Xinjiang. Managers and officials may wish to prepare for this possibility and/or take steps to prevent it from occurring.

The methodological advance in using the Information-theoretic approach for predicting the invasion of exotic plants is encouraging. Presented with the challenge of developing a model based on presence-only information, we developed a simple algorithm to produce a relative suitability map that makes reasonable predictions. We feel that an approach that uses logistic regression grounded in an information-theoretic framework offers a high degree of transparency desirable for explanation and understanding of ecological process as well as a robust platform for prediction. Future models might allow the prediction of a potentially invasive species based solely on its distribution in a native range before it arrives in a new region. Such ability would be highly desirable from the perspective of invasive species management. Before this point, however, it is important that we continue to evaluate this approach using a variety of species to ensure its robustness. We are currently doing just that using not only species native to North America and invasive in China, but using Chinese native species that are invasive to North America. Having international data and cooperation is a key asset in undertaking such works.

Acknowledgements

We gratefully acknowledge the invaluable assistance either through personal assistance or the provision of databased specimen records provided by the following herbaria: ALTA, ARIZ, AUA, F, JEPS, KUNHM, LL, MO, MONTU, MSC, NAU, NSPM, NY, OKL, PH, QUE, SASK, SUWS, TAMU, TEX, UA, UBC, UCR, UMO, UNB, USGB, UWPL, UWSP, WIS, WTU. In addition we acknowledge the herbarium museum of the Kunming Institute of Botany and the herbarium museum of the Institute of Botany of the Chinese Academy of Sciences.

The work was supported by the National Natural Science Foundation of China (No.40371084), U.S. Geological Survey (contract 03CRCN0001), and U.S. Geological Survey cooperative agreement (03CRAG0016). We wish to thank Drs Zhiliang Zhu and Qing-Feng Guo of USGS for their research collaboration, Professors Jun Chen and Yousong Zhao of NGCC for their constructive suggestions and comments, and M. Turner, D. Anderson, and other members of the Turner Laboratory for Ecosystem and Landscape Ecology for their helpful ideas and comments.

References

[1] Anderson DR and Burnham KP (2002) Avoiding pitfalls when using information-theoretic methods. Journal of Wildlife Management 66:912–918

[2] Anderson DR, Burnham KP, Thompson WL (2000) Null hypothesis testing: problems, prevalence, and an alternative. Journal of Wildlife Management 64:912–923

[3] Anderson DR, Burnham KP, White GC (2001) Kullback–Leibler information in resolving natural resource conflicts when definitive data exist. Wildlife Society Bulletin 29:1260–1270

[4] Austin MP (2002) Spatial prediction of species distribution: an interface between ecological theory and statistical modeling. Ecological Modeling 157:101–118

[5] Boulet LP, Turcotte H, Laprise C, Lavertu C, Bedard PM, Lavoie A, Hebert J (1997) Comparative degree and type of sensitization to common indoor and outdoor allergens in subjects with allergic rhinitis and/or asthma. Clin Exp Allergy 27:52–59

[6] Burnham KP and Anderson DR (2001) Kullback–Leibler information as a basis for strong inference in ecological studies. Wildlife Research 28:111–119

[7] Burnham, KP and Anderson, DR (2002) Model selection and multimodel inference. A Practical Information-Theoretic Approach. Springer, New York

[8] Cui LG, Shi DJ, Chen YM (2004) Survey and research for ragweed. Jilin Agriculture 178:20-21

[9] Creticos PS, Reed CE, Norman PS, Khoury MS, Adkinson NF, Buncher CR, Busse WW, Gadde J, Li JT et al. (1996) Ragweed Immunotherapy in adult asthma. J Allergy Clin Immunol 334:501–506

[10] Greaves RK, Sanderson RA, Rushton SP (2006) Predicting species occurrence using information-theoretic approaches and significance testing: An example of dormouse distribution in Cumbria, UK. Biological conservation 130: 239-250

[11] Guisan A and Zimmermann NE (2000) Predictive habitat distribution models in ecology. Ecol. Mode 135:147–186

[12] Guo QF (2002) Perspectives on trans-pacific biological invasion. Acta Phytoecologica Sinica 26:724-730

[13]Hyndman RJ and Fan Y (1996) Sample quantiles in statistical packages. American Statistician 50:361-365

[14]Kullback S and Leibler RA (1951) On information and sufficiency. Annals of Mathematical Statistics 22:79-86

[15]Legendre P and Legendre L (1998) Numerical ecology developments in environmental modeling, vol. 20. Elsevier, Amsterdam

[16]Loveland TR, Reed B C, Brown JF, Ohlen DO, Zhu Z, Yang L, Merchant JW (2000) Development of a global land cover characteristics database and IGBP discover from 1-km AVHRR data. International Journal of Remote Sensing 6:1303–1330

[17]Manly BF, MacDonald LL, McDonals TL, Thomas DL, Erickson, WP (2002) Resource selection by animals: Statistical design and analysis for field studies. Kluwer, Dordrecht

[18]Manel S, Dias JM, Ormerod SJ (1999) Comparing discriminant analysis, neural networks and logistic regression for predicting species distributions: as case study with a Himalayan river bird. Ecol. Model 120:337–347

[19]New M, Lister D, Hulme M, Makin I (2002) A high-resolution data set of surface climate over global land areas. Climate Research 21:1-25

[20]Osborne PE and Sua´rez-Seoane S (2002) Should data be partitioned spatially before building large-scale distribution models? Ecological Modeling 157:249-259

[21]Peterson AT (2003) Predicting the geography of species' invasions via ecological niche modeling. Quarterly Review of Biology 78:419-433

[22]Waller E and Zhu Z (1999) Estimating subpixel forest cover with 1-km satellite data. In: Proceedings of the Pecora14 Conference, Demonstrating the Value of Satellite Imagery, Denver, pp 419-426 (CD-ROM).

[23]Yoccoz Ng (1991) Use, overuse, and misuse of significance tests in evolutionary biology and ecology. Bulletin of the Ecological Society of America 72:106–111

An Evaluation of a GIS-aided Garbage Collection Service for the Eastern District of Tainan City

Jung-hong Hong and Yue-cyuan Deng

Department of Geomatics, National Cheng-Kung University, Tainan, Taiwan

Abstract: To provide citizens with a healthy living environment is a solemn duty of governments in the modern societies. The GIS technology opens new dimensionalities and possibilities to health-related research with its capabilities to record, manage and analyze spatio-temporal phenomena. Garbage collection is a common challenge to modern cities and efficiency is an important indicator of the dedication of the governments in this regard. With the so-called "no garbage on the ground" policy, garbage in the Tainan city has been collected by garbage trucks following fixed routes and schedules. It is therefore important for information about garbage collection to be widely distributed to the public and the designated routes and schedules to match the needs and living patterns of the residents. A web-based query system has been developed to allow for the search of accessible garbage collection points with given spatial and temporal constraints. The routes and schedules of garbage collection trucks were further analyzed using the four indicators we proposed and together with a street address database, to evaluate the efficiency of the garbage collection service. The results could serve as a basis to refine or adjust truck routes and schedules. In comparison with the traditional handbill approach and the adoption of a homepage of static contents, the GIS-aided approach clearly provides more thorough and powerful spatial analysis capabilities to addressing the garbage collection issue.

Keywords: garbage collection, GIS-aided analysis, serviceable area

1 Introduction

Developments made in the human civilization have dramatically revolutionized how we live in the world. With innovated technologies continuously emerging, we have indulged ourselves in the convenience and excitement modern technologies have brought us and abused recklessly limited resources and the ecological environment of this planet. One devastating phenomenon is the steadily growing volumes of garbage collected worldwide over the past years. Some garbage may contain chemical wastes that will have deteriorating effects to the environment for hundreds of years (Zheng, 2000).

Since garbage is an inevitable production of wastes from our daily lives, the critical challenges ahead of us are how to reduce volume, control pollution and make the best of its reusable portions. Some of these issues may be resolved only with technological innovations and huge capital investments in civic programs, but there are issues that can be improved by cooperation between the governing bodies and ordinary citizens like us. For example, the sorting of family garbage into recyclable and kitchen wastes can effectively reduce its volume and create profit making opportunities through the processing of recyclable items. With an average garbage volume per day in Taiwan reaching its historical highpoint of 24,331 tons in 1997, a remarkable 20% drop in the garbage volume to 19,876 tons in 2001 within a span of five years was achieved with the promotion of garbage sorting and recycling programs (Su, 2003). The citizens of Taiwan have helped create a better and healthier living environment with relatively little effort from every family. However, the success of the garbage reduction program certainly cannot depend solely on the introspective nature of the general public. Because garbage is collected and processed by the governments, the efficiency of the garbage collection service plays a crucial and indispensable role in achieving a better living environment.

For the perspective of an ordinary family, garbage may mean a few trash bags needing dumping. From the perspective of the government, garbage disposal involves such complicated issues as collection, cleaning, incineration and landfill. Among the 32 healthy cities indicators WHO (World Health Organization) proposes, the "Household waste collection quality index" and the "Household waste treatment quality index" are used to evaluate the progress on garbage issues (Hu and Tsai, 2004). Sponsored by the Bureau of Health Promotion (BHP), the Healthy City Research Centre (HCRC) of the National Cheng Kung University conducted a 3-year project to investigate and assess healthy city related issues in Taiwan.

The HCRC, upon choosing Tainan city as the research site, has maintained a close relationship with the Tainan city government and worked together to suggest a list of healthy cities indicators. This list of indicators included both international indicators from the WHO and local indicators to encompass distinguishing features of the Tainan City (HCRC, 2005). Among the proposed indicators, "Garbage volume per person per day" is designed to measure efforts incurred by residents and the city governments in reducing the garbage volume. The same indicator is also included in the

"Sustainable Development Indicator System" to evaluate Taiwan's progress in environmental protection (National Council for Sustainable Development, 2006). In addition, garbage volumes processed by landfill or incineration are included. Though these indicators have been widely used and often cited in the annual statistical reports of the city governments in the past, we must point out that they only reflect an overall measure of the garbage volume and fail to present a clear picture about the efficiency and effectiveness of the garbage collection service from the spatio-temporal perspective.

Back in the 1970s, garbage in Taiwan was collected by garbage trucks following fixed-routes and fixed-schedule, such that residents were required to wait at the garbage collection points for garbage trucks to collect their garbage. This approach was later modified to offer more flexibility by allowing residents to leave their garbage at the garbage collection points within an allowable period of time and not requiring them to wait for the garbage trucks to arrive. Though this approach was welcome by the residents, the accumulated garbage nonetheless caused environmental nuisance, triggered complaints from neighboring residents, and attracted unpleasant insects and animals. Many cities therefore began to promote the so-called "no garbage on the ground" policy about five years ago, which basically reverted to the old way of garbage collection. While the neighboring environments of garbage collection points were dramatically improved, the traditional approach to garbage collection meant inflexibility for the residents.

With a limited number of trucks and crews, an efficient dispatch plan of truck routes and garbage collection points can significantly reduce inconvenience to the residents. The current dispatching process, however, lacks a mechanism to quantitatively measure the efficiency of the garbage collection service as it relies on the dispatchers' experience and reactions from the general public to make adjustments. GISs have been successfully used, in the past 10 years, in various spatio-temporal domains including public health matters (Melnick, 2002; Soret et al., 2003). Since garbage collection is a typical problem with a spatial and temporal context, we believe the introduction of a GIS can bring new insights to the garbage collection issue. With the Internet technology, WebGISs have enabled a wider and easier access to geographic information by the general public. Location-based GISs, on the other hand, can instantly forward such information to public users. It is necessary for city governments to have an accurate and objective assessment about whether all households are well served and to identify regions with a poor service provision.

In this chapter, we focus on the use of a GIS toward (1) better dissemination of garbage collection information, and (2) the development of a quantitative measure of the quality of garbage collection service. The remaining parts of this chapter are organized as follows: section 2 discusses the evaluation model for the quality of garbage collection service, which leads to the design of four different indicators; section 3 demonstrates the WebGIS prototype we built for querying collection information and service evaluation; and section 4 presents some of our major findings and suggests possible future works.

2 Development of Evaluation Models

2.1 The current approach

Garbage collection service in the Tainan city is operated by The Bureau of Environmental Protection (BEP). In 2002, the BEP began to promote the "no garbage on the ground" policy. To help citizens gain a better understanding of accessible garbage collection services, every household was to receive a paper-based handbill that included the following information: (1) a map about the truck routes and garbage collection points in their neighborhoods, and (2) descriptive information about garbage collection points of the suggested routes, e.g., the order of stops, names of the neighborhood, the arrival and departure times, the duration of visit and location descriptions of the collection points with landmarks and/or street addresses (Figure. 1). Since contents of these handbills were not specifically customized for individual households, residents must read the information provided to determine the most convenient garbage collection points for them (by considering the distance to the collection points and the arrival times of garbage trucks). Should the garbage collection plan failed to meet their needs, residents must turn to the BEP's homepage to download route information of other trucks (i.e. an electronic version of other handbills) to make alternative choices. Unfortunately, the BEP's homepage is organized by individual truck route instead of regions or street addresses of households, so it is not a straightforward process for residents to find other garbage collection points near their homes. Let us suppose that a resident normally arrives at his/her home after 8pm, he/she has to first find all truck routes within the neighborhood and inspect the respective collection points one by one to determine an alternative collection point near the house with an arrival time of 8pm and beyond. This process can be very tedious if the resident has not sufficient spatial cognition about his/her neighborhood.

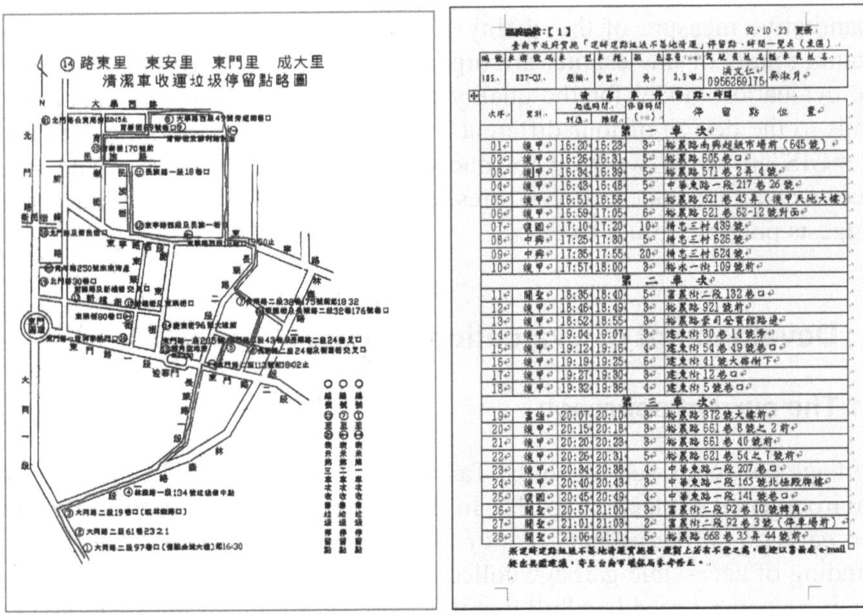

Fig.1. A printed handbill of garbage collection information

2.2 Factor Analysis

Whether residents are satisfied or not with their existing garbage collection service may largely depend on its “degree of convenience”, meaning that

- the location of garbage collection points must be well distributed such that residents do not need to walk very far; and
- the arrival times of the garbage trucks must match the living patterns of residents such that they are not away from home when the trucks arrive.

Since the location of garbage collection points, the arrival time of garbage trucks and the location of households are pre-determined, the search for qualified garbage collection points for a particular household can be typified as a constraint with both spatial (i.e., how far is the collection point from a residence) and temporal (i.e., will I be home when the garbage trucks arrive) components. Residents must specify an acceptable and reasonable distance and the preferred time period to successfully create a spatio-temporal constraint. For a garbage collection point to qualify as a candidate, its distance to the residence must be less than the given distance constraint and the arrival time of garbage trucks must be within the preferred time period. Based on this principle, three ideal scenarios can be identified:

1. At least one collection point is within a comfortable walking distance from the residence

Residents normally need to carry garbage and walk from their houses to the collection points, so the selection of an accessible collection point is determined by the distance. The choice of a "comfortable waking distance" is subject to an individual's physical condition and may differ from one person to another. Assuming the length of a normal step is between 0.6 and 0.7 meters and the normal walking speed is 100~120 steps per minute, the walking distance per minute is between 60 and 84 meters. If we further make an assumption that residents are only willing to walk for about 2 minutes at the most, "150 meters" appears to be an approximate and acceptable walking distance. We can then evaluate how many collection points are within the range of 150 meters from a particular household using geometric calculations. An ideal scenario is that every household has at least one accessible collection point within a given distance constraint.

2. The arrival times of garbage trucks match the living patterns of the residents

Based on their living patterns, residents will have their own preferred times for dumping garbage and will wish the arrival times of garbage trucks can meet their demands. In an urban area, the so-called "double income" families are very common. Since family members need to work in the daytime, their preferred service hours are often between dinner and bed times. On the other hand, business operations (e.g., stores) often have specific opening hours of 10am to 8pm and thus more flexible about the arrival times. The identification of communities with a high density of double-income families for designation of garbage collection in the evening hours can improve the quality of garbage collection service. An ideal scenario is that all households can find accessible collection points that match their temporal constraints.

3. All households are serviceable based on spatio-temporal constraints

By specifying the constraints of walking distance and arrival time, we can calculate the serviceable area of current garbage collection points and determine their serviceable households by testing the "within" relationship between the locations of serviceable area and the households. An ideal scenario is that all households are serviceable based on the residents' spatio-temporal constraints. The locations of the non-serviceable households represent the "weak regions" and the garbage collection plan must be adjusted accordingly. Both the number of non-serviceable households and their spatial distributions should be presented to the authorities for further reviews.

2.3 Indicator Design

Following the above discussion, an evaluation of the garbage collection service requires at least three types of fundamental data:

1. Garbage truck routes and collection points

Garbage trucks follow pre-designed routes and stay at collection points for a certain period of time according to the schedules. A truck route is geometrically modeled as a linear feature, while its associated collection points are modeled as an ordered set of point features (Figure 2). The attributes of a truck route include the route ID, type of truck, plate number, truck weight, the name of the driver, etc., while the attributes of a collection point mainly comprise the abovementioned descriptive information in the handbill. These two types of data share a common attribute named "route ID" that affords the query of route information from a single collection point and vice versa. Since the BEP did not own GIS-ready data of garbage truck routes and collection points, the required data were created manually from existing road network maps, street address databases and the distributed handbills.

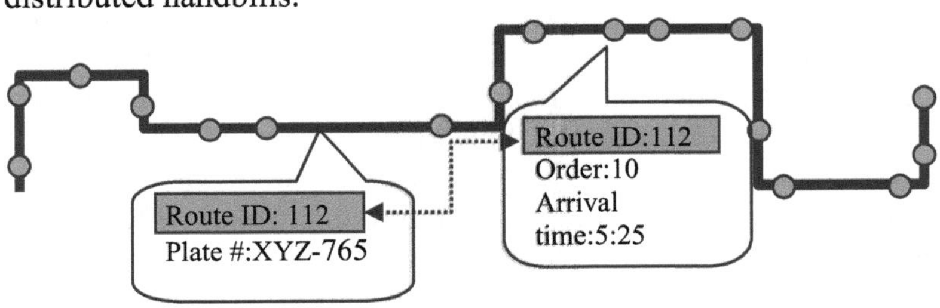

Fig. 2. A geometric depiction of truck routes and garbage collection points

2. Street address database

The street address database of Tainan city was created by a city-wide field survey founded on 1/1000 topographic maps. Every street address in the database represents a house and has a pair of associated plane coordinates (i.e., a point). For a multi-stories building, there may be a number of street addresses referring to the same coordinates. According to the survey specification, the recorded coordinates are preferably near the centre of a building, but this is not a rigorous requirement. Given the circumstance, the distance calculation for large buildings will invariably be less accurate. The street address database is constantly updated whenever necessary to ensure the completeness of all legal street addresses for our analysis.

3. Urban planning landuse district map data

Urban planning or landuse district data, which show land uses regulated by law, provide a good indication of the living patterns of households. This type of data is maintained by the Bureau of Urban Planning and Development. Each landuse district is represented as a polygon feature with its landuse as a major attribute. Each household is assigned a landuse value with the point-in-polygon test. Lands mainly designated for family housing are normally classified as the "building" landuse.

As mentioned earlier, whether a garbage collection point is regarded as a candidate or not is based on the result of the "within" test between the location of households and the serviceable area of that particular collection point. Theoretically, the delineation of serviceable areas should be based on the distance following existing roads (similar to the concept of "Manhattan" or "taxicab" distance discussed in Gartell, 1991) and not the Euclidean distance. However, because the current road data fail to provide all roads in the research area, we have to adopt a simplified approach of the Euclidean distance in this chapter. Assuming the location of both households and collection points are point features; the serviceable area is represented by a circle of a fixed distance using the well-known buffer operation (Figure 3). Should the calculated Euclidean distance between the residence and the collection point be less than the given distance constraint, the collection point would be regarded as a qualified collection point. Since the Euclidean distance represents the minimum distance between two point features, the analyzed result is naturally more optimistic than what would happen in reality. This limitation can only be improved with the availability of a more detailed road network data.

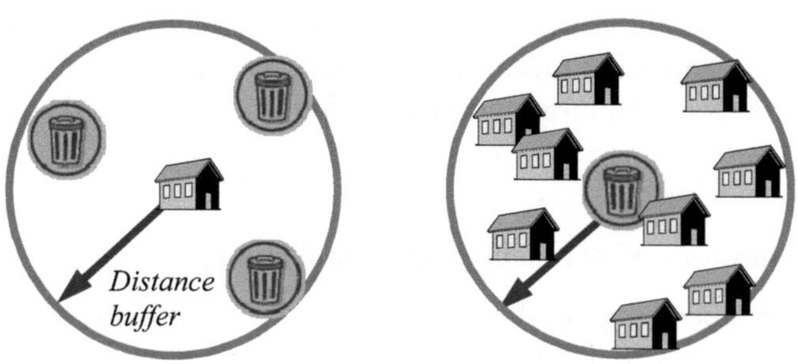

Fig. 3. The simplified distance buffer scenario for serviceable areas

Having determined the serviceable areas of all collection points, we can spatially aggregate these serviceable areas into a "total serviceable area" representing the current configuration of collection points. Depending on its location, an individual household may have zero accessible collection point (i.e. outside the serviceable area) or one to many accessible collection points. Figure 4 illustrates this scenario in which the triangle symbols represent individual households and their associated labels denote the number of their accessible collection points. Given two truck routes in this figure, an individual household may have access to collection points along both routes and sometimes their arrival times may be very different (e.g., 4pm and 8pm) to provide more flexibility.

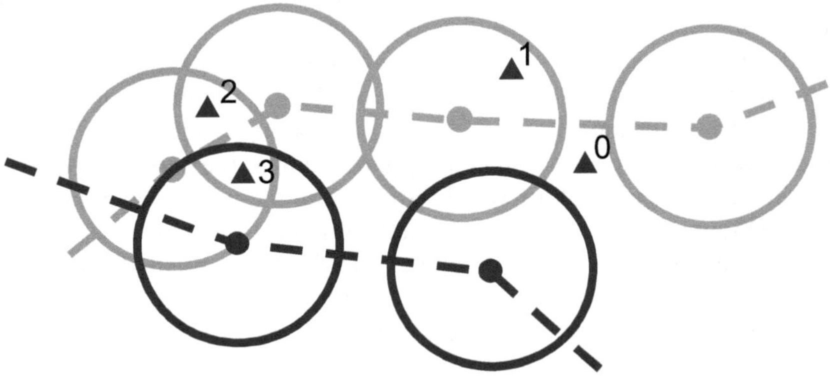

Fig. 4. Number of serviceable collection points by households

While past indicators have concentrated on garbage volume only, we argue that the quantitative measure of garbage service quality should present a more precise spatio-temporal meaning and propose the following four indicators.

1. The total number of non-serviceable households

This indicator reports the total number of households "outside" the "total serviceable area" based on existing configuration of garbage collection points and the use of spatial/temporal constraints. The spatial distribution of non-serviceable households shows the "weak regions" with poor services such that the residents are required to walk farther than expected to dump their garbage. Were the distance buffer to increase because residents are willing to walk farther, the value of this indicator should decline. But for an ill-distributed configuration of garbage collection points, the improvement may not be significant. Under such a circumstance, the authorities must further inspect the distribution of these non-serviceable households to make necessary adjustments to the truck routes and locations of garbage collection points.

2. The number of accessible collection points for individual households

With a given distance constraint, this indicator records the total number of accessible collection points for individual households regardless of their arrival time. After determining the number of accessible collection points for every household, the analyzed result can be further categorized by the number of accessible collection points. The total number of households for each category can also be used to measure the service coverage.

3. The total number of non-serviceable households after 6:30 pm

With a given walking distance constraint, this indicator shows the number of non-serviceable households after 6:30pm. As the off-duty time in Taiwan is usually between 5pm and 5:30pm, we assume that most residents would be back at home before 6:30pm. This indicator serves as a quantitative measure about the overall provision of garbage collection service at night which is particularly important to households located in the "building" landuse districts. Given that "6:30pm" is an assumption in this chapter, the time can of course be adjusted accordingly.

4. The total number of households serviceable both before and after 6:30 pm

This indicator reveals the total number of households fulfilling the following conditions: (i) multiple accessible collection points, (ii) at least one collection point whose arrival time is before 6:30pm, (iii) at least one collection point whose arrival time is after 6:30pm, and (iv) the duration between two arrival times is more than 30 minutes. This indicator demonstrates an ideal service scenario to individual households, as residents are provided more flexibility as far as the arrival time is concerned.

3 Analyses and Results

To demonstrate the proposed evaluation mechanism, the "eastern district" of Tainan city was chosen as our research site (Figure 5.a). Tainan city with a 2005 population of more than 700,000 is one of three early-developed areas in Taiwan in the 18th century and still one of two major cities in southern Taiwan. Since lands in the downtown area have been fully developed, new developments have turned to the suburb districts over the recent 20 years. The eastern district is the fastest developed district in terms of population growth. The 2005 census data reported that the "eastern district" now owns the most population (25%) among the six districts of the Tainan city. About half of the lands in the eastern district belong to the "building landuse" (Figure 5.b). In this chapter, we created 25 garbage truck routes and 895 garbage collection points based on the information collected from the BEP. It was documented in the Tainan city government's annual report that the average garbage cleaning volume per day exceeded 600 tons in the late 90', but has since gradually declined to about 580 tons in 2005.

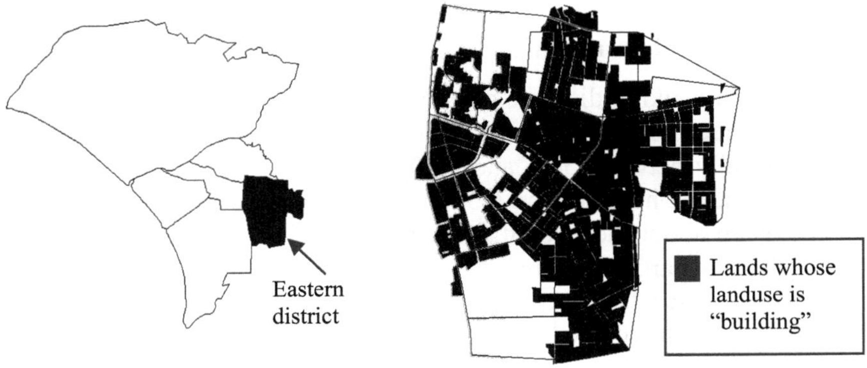

Fig. 5. (a) Eastern district of Tainan (b) Lands of the "Building landuse" type in the "Eastern district"

3.1. A Web-based Garbage Information Query System

One apparent shortcoming of the current approach is that only information about selected truck routes is presented to residents and this information unfortunately is not customized on the basis of an individual's needs. By effectively integrating all related data into a GIS, the additional query capabilities will enable a resident to search qualified collection points within the given spatial and temporal constraints that suit the individual's need. Furthermore, the search will not be limited to only the suggested truck routes or nearby collection points. The query system provides the necessary flexibility and most importantly, the queried result is totally based on individual preference and is therefore customized. To overcome limitations of the current static approach, the eventual query system should provide the following capabilities:

- Easy access by the general public
- Ability to integrate relevant and related data
- Ability to allow an individual to specify spatial and temporal constraints
- Ability to filter alternative garbage collection points within given constraints
- Ability to map the locations and display attributes of collection points

It is obvious that the traditional handbill approach cannot be relied on if a wider public access to updated garbage collection information is to be realized. Given the fact that households in Taiwan having connectivity to the Internet reached 50% in 2003 (TWNIC, 2003) and is expected to grow with the completion of wireless mobile networks in the coming years, the WebGIS appears a good alternative for wider and easier information dissemination.

A web-based garbage information query system has been developed with the AutoDesk MapGuide. The interface allows a user to specify online the location of his/her houses (with a street address or via a map click) and the distance/temporal constraints (e.g., Find collection points within 100 meters from No.77, Shin-Li Road and whose arrival time is between 7am and 10pm) (Figure 6). Street address is the default input method for specifying a residential location. The online query process will return a map showing the specified street address and its neighboring environment, plus highlighted symbols indicating the locations of qualified garbage collection points (Figure 7). The arrival times and location description of garbage collection points are also displayed on the side to permit further inspection. In comparison with the traditional handbill approach, a distinct difference for the residents is the capability to customize spatial and temporal constraints and find collection points targeted for individual needs without extraneous information. Furthermore, whenever truck routes and schedules have been modified, the WebGIS system can immediately offer updated information to the residents, which is practically impossible under the traditional approach.

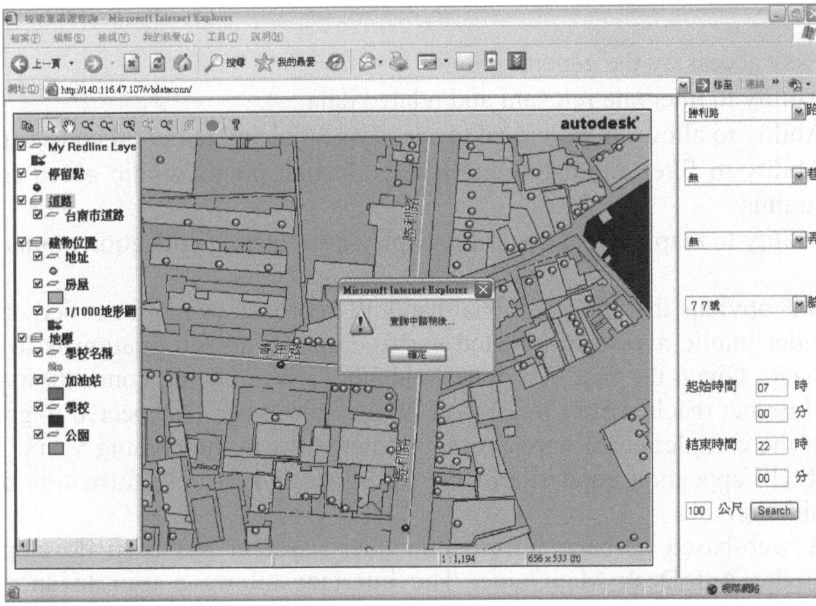

Fig. 6. The map interface for querying garbage collection points with a street address

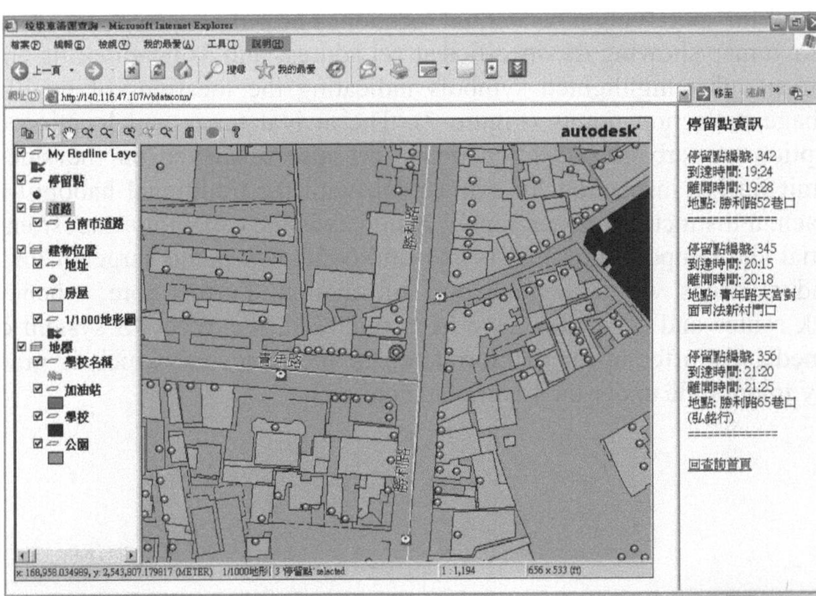

Fig. 7. A queried result of garbage collection points

3.2. An Evaluation of the Garbage Collection Service

There were altogether 71,246 households in the research area based on the street address database provided by the city government. We analyzed the serviceable areas for 895 garbage collection points using five different distance constraints: 50 meters, 75 meters, 100 meters, 125 meters and 150 meters. For every chosen distance constraint, we can determine the number of accessible collection points for individual households and calculate the total number of households on the basis of the number of accessible collection points (Table 1). For example, under the column of "50 meter" buffer, "25,236" is interpreted as "there exists 25,236 households with zero accessible collection points within 50 meters of their residence." Similarly, 95 households have access to 4 garbage collection points. Some other findings are discussed as follows:

1. It can be seen from Table 1 that the number of non-serviceable households decreased from 25,236 (35% of the total households) to 62 households (0.08%) with increasing buffer distances from 50 meters to 150 meters. Indeed, only about 2% of the households would remain non-serviceable were the residents willing to walk 100 meters to the collection points.

Table 1. Statistics of the number of accessible collection points and the number of households based on different distance buffers

Buffer / Number of accessible collection points	50 m	75m	100m	125m	150m
0	25236	6894	1319	197	62
1	33494	24508	8834	3173	1117
2	11105	22569	16230	6907	3578
3	1496	12324	18812	9486	3505
4	95	3822	13501	14820	7258
5		1195	7989	12390	9246
6		103	3279	10438	11023
7		11	1067	7737	9989
8			311	3842	8071
9			75	1572	7005
10			6	622	4647
11			3	191	2901
12				31	2008
13				20	734
14					165
15					104
16					13

2. With the 100 meter distance buffer, 86% of the households would have access to more than one collection points in their neighborhood. The upper curve in Figure 8 shows that the number of serviceable households did not improve significantly beyond the 100 meter distance buffer. If "100 meters" was an accepted walking distance, we could reasonably argue that most of the households would be well served with the current configuration of garbage collection points and the general public should be satisfied with the service provision. On the contrary, if "50 meters" was the acceptable walking distance, then the result would not be encouraging and modification to the garbage collection service would be necessary.
3. As the distance buffer increases, the number of accessible collection points for individual households also increases. About 90% of the households would have more than 3 accessible collection points with a distance buffer of 150 meters. This finding is very interesting because residents tended to accept collection points suggested in the handbills and adapt individual schedules to fit the schedules of garbage trucks, but the analyzed result indicates a few more alternatives to the current options.

We are particularly interested in analyzing the provision of nighttime garbage collection service (i.e., services after 6:30pm). The lower curve in Figure 8 shows the number of serviceable households with after 6:30pm services for different buffer distances. Table 2 shows more than 50% of the households were not within serviceable area if the distance buffer was 50 meters and the number of serviceable households would increase with larger distance buffers. Nevertheless, about 5000 non-serviceable households would still remain even with the 150 meter distance buffer.

Table 2. Number of non-serviceable households after 6:30 PM

Buffer	# of non-serviceable households
50 m	38251
75 m	23679
100 m	14588
125 m	8583
150 m	4992

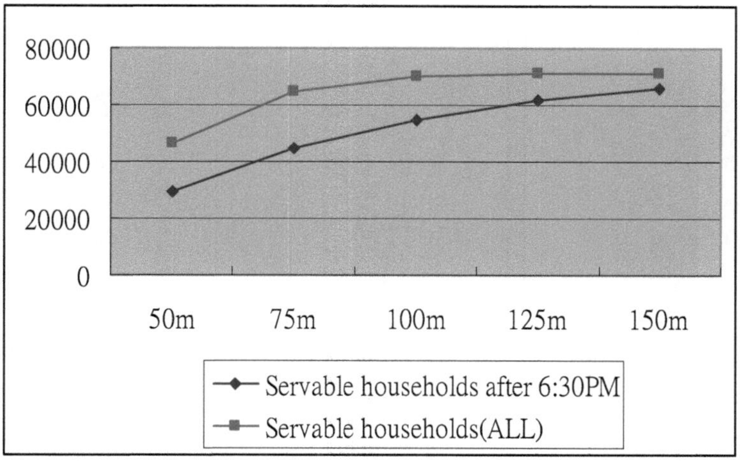

Fig. 8. Number of serviceable households with different conditions

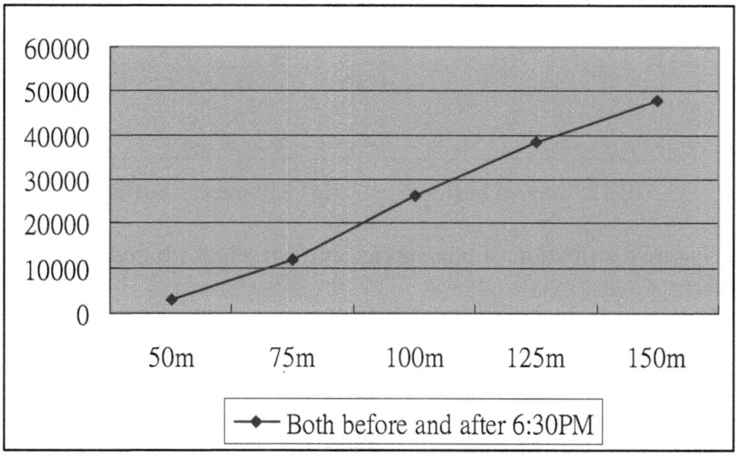

Fig.9. Number of households serviceable both before and after 6:30pm

Figure 9 shows the number of households serviceable both before and after 6:30pm. This quantitative measure allows for an assessment of the provision of garbage collection services. Even with the number of serviceable households steadily increasing as the distance buffer increases from 50 meters to 150 meters, only 26,125 qualified households (about 1/3 of the total households) are served at the 100 meter distance buffer. Unless a higher level of garbage collection service is expected both before and after 6:30pm by the majority of residents, the current schedule and configuration of garbage collection points need no major modification.

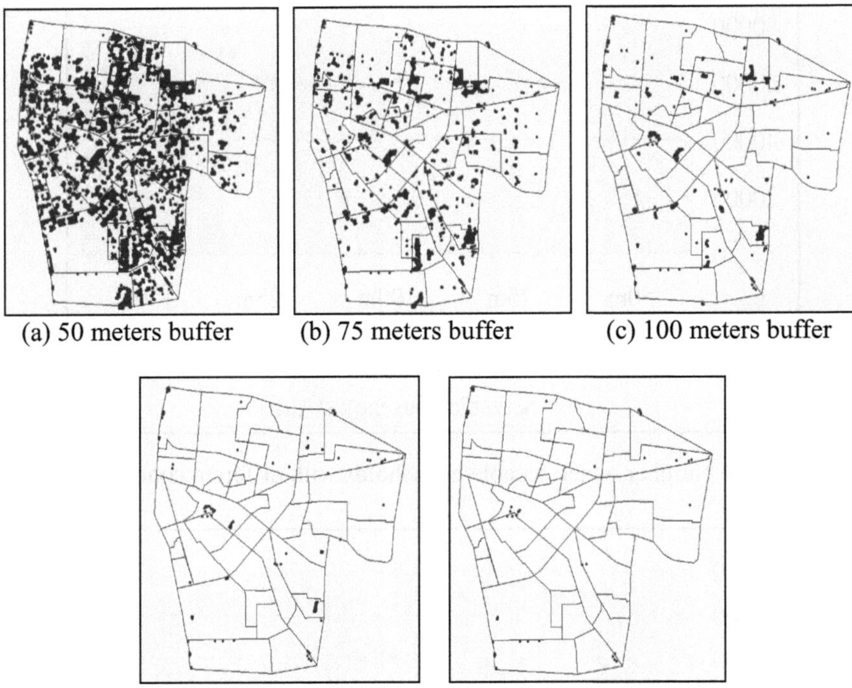

(a) 50 meters buffer (b) 75 meters buffer (c) 100 meters buffer

(d) 125 meters buffer (e) 150 meters buffer

Fig. 10. Spatial distribution of non-serviceable households based on different distance buffers

The above indicators provide overall measures about the "quality" of garbage collection service. To improve the garbage collection service, the most important thing is to identify non-serviceable households and adjust collection routes and schedules accordingly. Figures 10 (a) - (e) illustrate the spatial distribution of non-serviceable households assuming different distance buffers. A 50 meter distance buffer yielded a widespread distribution of non-serviceable households which did not seem realistic. As the distance buffer was increased, it became easier to identify "weak regions" with non-serviceable households. Such maps are potentially useful in identifying areas needing more collection points. It should be noted that some non-serviceable households are located close to the boundary of the eastern district, so they may be able to access garbage collection services provided by other districts. Figure 11 uses the thematic mapping technique to illustrate the number of accessible collection points for individual households based on a distance buffer of 100 meters; the darker symbols indicate the more accessible collection points a household has. From this figure, the authorities can see visually the levels of service provision.

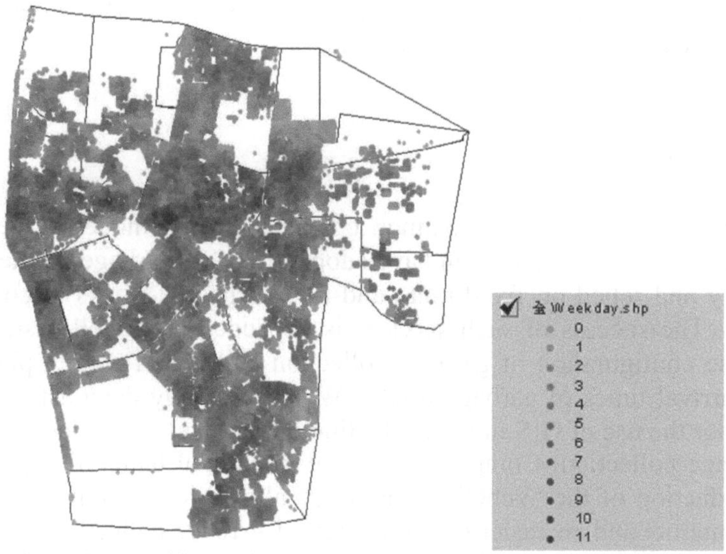

Fig. 11. A thematic map illustration based on the number of accessible collection points of individual households.

(a) 50 meters buffer (b) 100 meters buffer (c) 150 meters buffer

Fig. 12. A spatial distribution of non-serviceable households within the "building" landuse districts

A critical factor for evaluating the garbage collection service for households located within the "building" landuse districts is its level of nighttime service provision. Figures 12 (a) - (c) illustrate the spatial distribution of non-serviceable households after 6:30pm using three different distance buffers. When the distance buffer is 50 meters, it is clear that the provision level is rather poor. As the distance buffer increases, we can find a trend of spatial distribution similar to that of Figure 10. Since the non-serviceable households are again located within a limited area, the addition of more garbage collection points near these regions should be able to improve effectively the non-serviceable situations.

4 Conclusion

Garbage processing is a common challenge facing modern cities worldwide. Major cities in Taiwan have adopted the "no garbage on the ground" policy and relied on fixed-route and fixed-schedule trucks to collect garbage. The success of such services is depended on a well-designed geometric configuration of garbage collection points and careful planning of the arrival times of garbage trucks. We successfully demonstrated in this chapter the use of GIS in the evaluation of the level of service provision in garbage collection. Compared with the traditional handbill approach, the introduction of the WebGIS technology allows for ubiquitous access via the Internet and an easier dissemination of garbage collection information. Its flexible query capability further allows the search of all serviceable garbage collection points within a selected neighborhood.

We further proposed four indicators to quantitatively measure the garbage collection service based on the geometric distribution of households and collection points. The use of a GIS offers powerful tools to calculate indicator values and to undertake thematic mapping to illustrate the spatio-temporal nature of the analyzed results. Our analyses indicated that the proposed four indicators can highlight characteristics of a specific configuration of garbage collection points, particularly in reference to requirements of the residents. Our example of the eastern district suggested that the current configuration of collection points can provide sufficient services to its residents at a distance buffer of 100 meters. Because the locations of non-serviceable households are displayed on maps, the authorities are in a better position to adjust the routes and collection points to improve the service level of the "weak regions." For the garbage processing authorities, a GIS is therefore not only an invaluable aid to bridging the gap between the government and the general public, but also a powerful tool to improving the level of service provision in garbage collection.

Acknowledgements

The authors would like to express their appreciation to the Healthy City Research Centre of the National Cheng-Kung University for their assistance in the preparation of this chapter. Some of the ideas and technology used in this chapter are from the research result of an NSC sponsored research project (NSC94-2211-E-006-071).

References

[1] Gatrell AC (1991) Concepts of Space and Geographic Data, In: Maguire DJ, Goodchild MF, Rhind, DW (eds.) Geographical Information Systems, Volume 1: Principles, Longman Scientific & Technical, pp. 119-134
[2] Healthy City Research Centre (2005) White Paper for the Healthy City of the Tainan City, National Cheng Kung University in Tainan, 116 pages
[3] Hu SC and Tsai SY (2004) Concepts of WHO Healthy Cities, Journal of Healthy Cities, Taiwan, 1(1): 1-7
[4] Melnick AL (2003) Introduction to Geographic Information Systems in Public Health. Aspen Publishers. Inc. 300 pages
[5] National Council for Sustainable Development (2006) Sustainable Development Indicator System, http://www.cepd.gov.tw/sustainable-development/eng/main.htm, Accessed on March 10, 2006
[6] Soret S, McCleary K, Rivers PA, Montgomery SB, Wiafe SA (2002) Understanding Health Disparities through Geographic Information Systems. In: W. Khan (ed.) Geographic Information Systems and Health Applications, Idea Group Publishing, pp. 12-42
[7] Su HY (2003) The Garbage Sampling and Analysis in Taiwan in 2002: Final Report, http://ivy2.epa.gov.tw/out_web/twdep/www/d40/%E6%96%B0%E8%B3%87%E6%96%99%E5%A4%BE/Chapter_2.pdf, Accessed on January 15, 2006
[8] TWNIC (2003) Broadband Network Use Survey Analysis in Taiwan, Taiwan Network Information Centre, http://www.twnic.net.tw/file/TWNIC-Borband Survey-92.08.19.pdf, Accessed on March 10, 2006
[9] Zheng YM (2000) "Current Status of Garbage Processing and Questions in Taiwan." http://eec.kta.org.tw/incinerator/diox111.htm, Accessed on March 15, 2006

A Study of Air Quality Impacts on Upper Respiratory Tract Diseases

Huey-hong Hsieh[a], Bing-fang Hwang[b], Shin-jen Cheng[a] and Yu-ming Wang[c]

[a] Department of Environment and Resources Engineering, Diwan College of Management, Taiwan
[b] School and Graduate Institute of Occupational Safety and Health College of Public Health, China Medical University, Taiwan
[c] Department of Information Management, Diwan College of Management, Taiwan

Abstract: Upper respiratory tract (URT) diseases are highly related to air quality. Recent interests have focused on traffic-related air pollution and the potential health effects associated with such exposure. These studies have indicated a strong correlation between air pollution and its related diseases. With advances made in computer technologies, Geographic Information Systems (GIS) nowadays provide useful spatial analysis tools to examine spatial characteristics of air quality. Maps of the spatial distribution of air pollutants can be produced with a GIS to facilitate in-depth studies. Our study incorporated the 2003 Cancer Death Rate (CDR) map provided by the Bureau of National Health Insurance with the air quality data provided by the Environmental Protection Agency to examine the associations between air pollution and upper respiratory tract diseases. First, maps of select air pollutants (NO_X, O_3, PM_{10}, SO_2 and CO) were produced using the inverse distance weighting interpolation method. Then, a series of statistical analyses were performed to examine the correlation between these pollutants. Finally, a multivariate regression analysis of association was preformed to study the correlation between air pollution and URT diseases. The results indicated a strong correlation between NO_X and CO (coefficient of correlation 0.92), and association between these pollutants and URT diseases.

Keywords: upper respiratory tract disease, air quality, multivariate regression

1 Introduction

Recent research from the U.S., Netherlands and Australia reported an association between URT diseases and the air quality (Woolcock, et, al, 1997, Boezen, et al., 1999, Brunekreef, et al., 2002). In Taiwan, the relationship between the diseases and air quality has not yet been discussed. Air pollutant exposure on an individual can be estimated from measurements recorded by nearby monitoring gauges. However, air pollutants vary in space and time and the gauge measurements only provide point-based exposure values at an instance. To extend the point-based measurements of air pollutant exposure, spatial interpolation was applied on these point measurements using a GIS to estimate a pollution surface. Associative analyses were then applied to find the relationships between the URT diseases and the derived air pollutant values.

The procedures can help extract the most significant influential factors of URT diseases. In addition, the derived air pollutant maps can provide useful information for related analyses.

2 Data Characteristics

Three types of data sources were used for our analyses: (1) Air Quality System (AQS) database from the Environment Protection Administration (EPA), (2) Census database from the Department of Statistics of the Ministry of Interior (DSMI), and (3) the Cancer Distribution (CD) database provided by the Bureau of Health Promotion of the Department of Health (BHPDH). The following sections describe in greater detail the characteristics of each type of data.

2.1 Air Quality System (AQS) database

The AQS database contains measurements of air pollutant concentrations in Taiwan under the provisions of the Clean Air Act intended to monitor the quality of air. The EPA sets limits on the acceptable amounts of a range of pollutants present in the air. Included in the measurements are the five criteria air pollutants: Carbon Monoxide (CO), Nitrogen Dioxide (NO_2), Sulphur Dioxide (SO_2), Ozone (O_3), and Particulate Matter (PM_{10}). The air quality monitoring network of Taiwan consists of 58 gauges as illustrated in Figure 1. The measurements of each pollutant were interpolated into a pollution surface for further analyses.

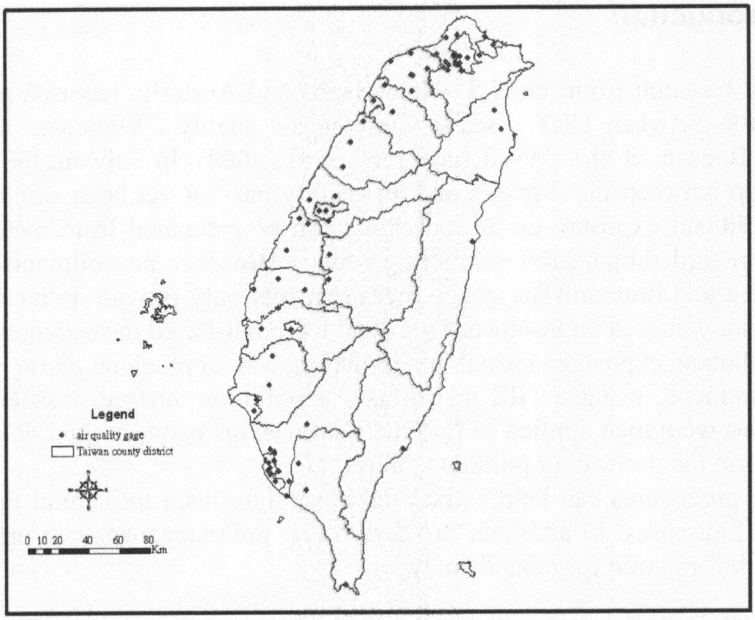

Fig. 1. The air quality monitoring network system in Taiwan (Source: Adapted from EPA, Taiwan)

2.2 Census database

The Census database provided by the DMSI contains monthly population number, birth rate and death rate, and other socio-economic data of each city. The population density of each county was derived by dividing its population by its area and was a major factor for further analyses.

2.3 Cancer Distribution (CD) database

The CD database provided by the BHPDH contains death rates of each city categorized by each disease. The URT cancer death rate of each city was used to conduct associative analyses with the air pollutants. The 1992-2001 cancer death rate map is as shown in Figure 2.

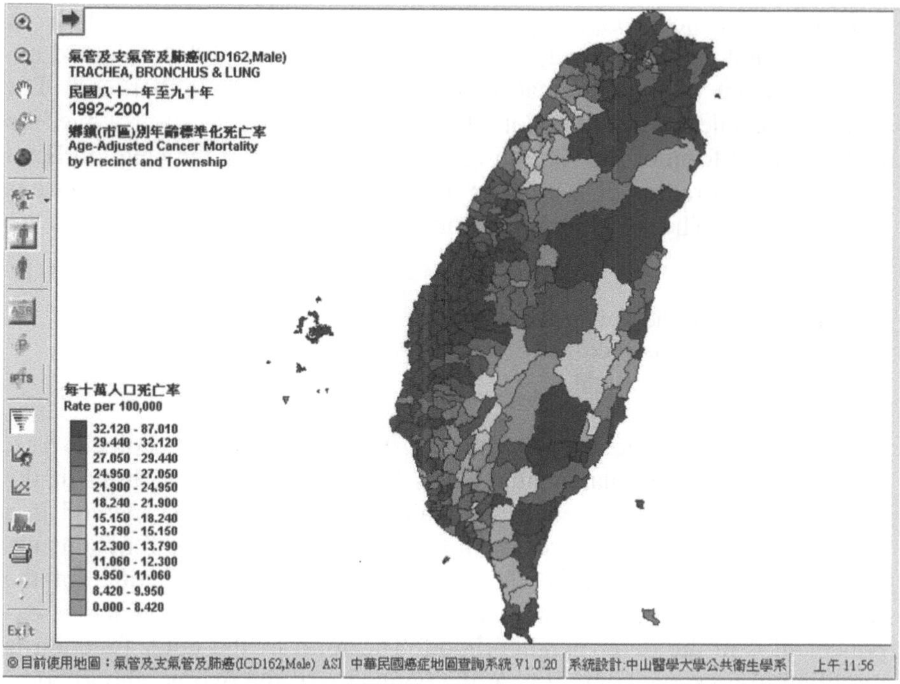

Fig. 2. Age-adjusted cancer mortality by precinct and township of Taiwan (Source: Adapted from BHPDH, Taiwan)

3 Methodology

As introduced in the previous section, three types of data were used in this study. Each data set was pre-processed for further analyses. Four procedures were involved in the analyses: (1) interpolation of the point data, (2) extraction of estimated values at specific locations on a surface map, (3) analyses of the correlation between air pollutants, and (4) association analyses between the URT diseases and air pollutants. The following sections briefly describe the processes.

3.1 Interpolation of point data

The air pollutant measurements from AQS database were integrated into monthly and yearly point data and interpolated to pollutant surfaces using the inverse distance weighting method. The method is briefly described as below.

Inverse distance weighting is the simplest interpolation method. A neighborhood about the interpolated point is identified and a weighted average is taken of the observation values within this neighborhood. The weights are a decreasing function of distance. The user has control over the mathematical form of the weighting function, the size of the neighborhood expressed as a radius or a number of points, in addition to other options (Fisher, 1987). The weighting function is formulated as follows:

$$w(d)= 1/d^p \qquad \text{Eqn. (1)}$$

where $w(d)$ is the weighted distance;
d is the distance; and
$p>0$ is the weight

The value of p is specified by the user. The most common choice is $p=2$. When $p=1$, the interpolated function is "cone-like" in the vicinity of the data points, where it is not differentiable. Shepard's (1968) method is a variation of the inverse power, with two different weighting functions using two separate neighborhoods. The default weighting function for the Shepard's method is an exponent of 2 in the inner neighborhood and an exponent of 4 in the outer neighborhood. The form of the outer function is modified to preserve continuity at the boundary of the two neighborhoods. The neighborhood size determines how many points are included in the inverse distance weighting. The neighborhood size can be specified in terms of its radius (in km), a number of points, or a combination of the two. The use of a radius requires the user to specify an override value in terms of a minimum and/or maximum number of points. Invoking the override option will expand or contract the circle as desired. Conversely, specification by number of points requires an override value in terms of minimum and/or maximum radius. It is also possible to specify an average radius as an override to expand or contract the size of a neighborhood.

3.2 Extraction of estimated values at specific locations on a surface map

The air pollutant surfaces derived from the previous section summarize the spatial distribution of each pollutant. To examine correlations among the pollutants and possible association between these air pollutants and the URT diseases, air pollutant readings at each URT disease locations were extracted from the derived concentration surface maps. The ArcGIS Spatial Analyst tool (developed by ESRI) was used to extract pollutant readings at specified locations on a surface.

3.3 Analyses of correlation between air pollutants

The correlation between a pair of air pollutants was calculated using the Pearsons coefficient of correlation (Edwards, 1976) expressed as follows:

$$R = \frac{n\sum xy - \sum x \sum y}{\sqrt{\left[n\sum x^2 - \left(\sum x\right)^2\right]\left[n\sum y^2 - \left(\sum y\right)^2\right]}} \qquad \text{Eqn. (2)}$$

where R is the correlation coefficient;
x and y are readings for pollutant x and pollutant y respectively; and
n is the number of measurements.

The correlation coefficient is between -1 and 1; with –1 and 1 indicating perfect negative and positive correlation between variables in the sample data and with a value of 0 indicating no relationship between the variables in the sample data.

3.4 Association analysis between URT diseases and air pollutants

The relationships between the URT diseases and the air pollutants can be assessed using the Kendall coefficient of concordance (Gibbons, 1997) which indicates the degree of association between the variables in the sample data. The test variables in ordinal measurement must be rearranged before the Kendall coefficient of concordance can be calculated. The following sub-sections describe briefly the procedures.

3.4.1 Data ranking

Suppose the data consist of a random sample of n pairs of observations with k variables, (X11, X21, X31, …, XK1), (X12, X22, X32, …, XK2), …, (X1n, X2n, X3n, …, XKn), where each set (that is X1 set, X2 set, …, XK set) is measured on an ordinal scale. For the purpose of the analysis, observations in each set are ranked. That is, the integers 1,2,3, …, n are assigned to X11, X12, X13 according to their relative magnitudes within the set of X1 observations. Let U1i denotes the rank of X1i, so the set U11, U12, …, U1n represents some arrangement of the first n positive integers. Similarly, the series of Xk1, Xk2, Xk3, …, Xkn is ranked; the results of this ranking are denoted by Uk1, Uk2, …, Ukn. Now the data to be analyzed consist of n pairs of positive integers (U11, U21, …, Uk1), …, (U1n, U2n, …, Ukn), which represent the pairs of ranks of the original observations. In the discussion to follow, we assume for convenience that these pairs are listed in increasing order of the value of the Xk ranks; that is Uki=I, for i=1,2, …, n.

3.4.2 Rationale for the descriptive measure

The rankings of the k characteristics are in perfect agreement if and only if U1i=U2i= … =Uki for each i. The rankings of the k data characteristics are in perfect disagreement if they are paired in completely reverse orders.

3.4.3 Kendall coefficient of concordance for complete rankings

As mentioned in section 3.4.1, suppose the data consist of k complete sets of rankings of n objects. Each ranking is made by a comparison of objects within the set only. The data as collected or observed consist of nk ranks or measurements on at least an ordinal scale. The rankings are assigned separately within each of the k group. The association between the k sets of rankings is of interest. For simplicity, the ranks of each set are assigned accordingly. The sum of ranks from X11, X12, …, X1n to Xk1, Xk2, …, Xkn are calculated accordingly. We use S1, S2, …, Sk to denote the sum of ranks for each set. The sum of ranks of each data set is also calculated and is summarized in total Sk. Let Sa denote the average of a perfect agreement of data sets, then Sa=Sk/n. If data sets are in perfect agreement, then S1=S2=… =Sk. Let S denotes the sum of squares of deviations between the observed sums and the expected sums, as expressed in the following form:

$$S=(S1-Sa)^2+(S2-Sa)^2+ \dots + (Sx-Sa)^2$$

If the k sets of rankings are in perfect agreement, the rank of X11, X12, …, X1n will be assigned rank 1 and the sum will be n. The sum of ranks of X21, X22, …, X2n will have a value of 2n, and so forth. We can then calculate the sum Sd of squares of deviations of these column sums (with perfect agreement) from the column expected (with no association):

$$Sd=(n-S)^2+(2n-S)^2+\dots+(n2-S)^3=\Sigma[Sj-k(n+1)/2]^2 \qquad \text{Eqn. (3)}$$

where j=1,2, …, n

The relative agreement can be measured by a ratio of the actual sum of squares to the sum of squares under a perfect agreement, or

$$W=Sd/Sa=12Sd/k^2n(n^2-1)$$

where W denotes the Kendall coefficient of concordance.

The ratio will clearly compute to 1 for a perfect agreement and tend toward 0 for no agreement.

3.4.5 Test of association

Given that the k sets of rankings result from measurements on k characteristics, the Kendall coefficient of concordance can be used to test the null hypothesis that these characteristics are independent. In other words, there is no association or relationship between the k variables. The hypothesis is set as below:

H_O: No association H_A: Association exists

The p-value or significance level is the right tail and can be determined from a reference table of the Kendall coefficient of concordance. A 95 percent confidence interval or 0.05 significance is used for the test.

3.5 Analysis procedures

The analysis procedures are outlined in the following steps:

Step 1: Find the URT death rate for each city from the CD database

Step 2: Transfer the death rate into a geodatabase

Step 3: Interpolate the point-based URT death rate data into a surface

Step 4: Extract from the surface derived in step 3 death rate values at the EPA questionnaire locations

Step 5: Interpolate the point-based air pollutant data of CO, NO_x, SO_2, O_3, PM_{10} into surfaces

Step 6: Extract from surfaces derived in step 5 values of pollutants at the EPA questionnaire locations

Step 7: Find population density values at the EPA questionnaire locations

Step 8: Examine correlation between air pollutants

Step 9: Examine correlation between URT death rate and population density

Step 10: Test for association between URT death rate and air pollutant values

4 Results

The analyses were performed following the procedures mentioned in section 3 and the results presented below.

4.1 Spatial distribution of air pollutants

The study first computed the 1996 yearly concentration of air pollutants from the AQS database. These point-based data were then interpolated into concentration surfaces as shown in Figure 3. Note that the mountainous area in the middle was removed from the analysis since there were no monitoring gauges in this area.

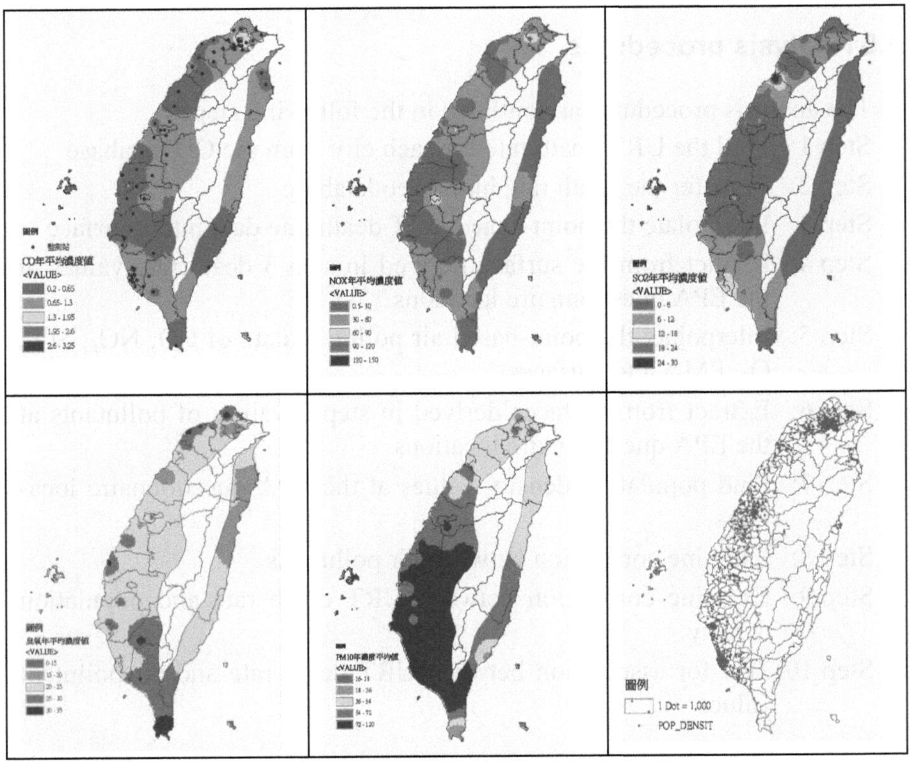

Fig. 3. 1996 yearly air pollutant distribution and population density (figures from upper left to right, then lower left to right show the concentration of CO, NO_x, SO_2, O_3, PM_{10}, and population density)

From the series of surface maps, we could see that pollution was more serious in the urban areas such as Taipei and Kaoshung. Southern Taiwan, however, registered a more serious level of air particulate suspension of PM_{10} than the north, possibly a result of uneven rainy days. In the northern part, the rainy season was longer than the central and southern parts of Taiwan during the study period and rain was suspected to wash away suspended particulates.

Table 1. Correlation coefficients between air pollutants

pollutant	NO_x	O_3	PM_{10}	SO_2	CO
NO_x	1.00	-0.47	0.04	0.32	0.92
O_3	-0.47	1.00	0.26	-0.11	-0.47
PM_{10}	0.04	0.26	1.00	0.51	0.26
SO_2	0.32	-0.11	0.51	1.00	0.51
CO	0.92	-0.47	0.26	0.51	1.00

The correlation between the various air pollutants were calculated and shown in Table 1 which shows a strong relationship between NOx and CO with a correlation coefficient of 0.92. Hence, one of the two pollutants needed to be excluded from the associative analysis between URT death rate and air pollutants. In the study, CO was removed from the analyses of association. Also SO_2 and CO are correlated ($R = 0.51$) and there exists a reverse correlation between NO_x and O_3 ($R = -0.47$).

4.3 Correlation between URT death rate and air pollutants

The correlation coefficient of the URT death rate and population density was 0.1, which indicates a weak relationship between them. The correlation coefficients between the URT death rate and NOx, O_3, PM_{10}, SO_2 and CO were: 0.26, -0.09, 0.03, 0.26, 0.25 respectively. These results indicate positive although weak relationships between the URT death rate and NOx, SO_2 and CO.

4.4 Associative analysis

Given the weak correlation between the URT death rate and NO_x, SO_2 and CO, these three pollutants were taken into further consideration for associative analysis with the URT death rate.

The test was performed in SPSS and the test statistics was 0.9 with a p value of near 0 indicating rejection of the null hypothesis. This result implies that a strong association of the URT diseases and the pollutants.

5 Summary and suggestions

This study combined three types of data in an analysis of pollutants and their possible effects on the URT diseases. Point-based data on pollutants were used to derive air pollutant surfaces which facilitated the study of the relationships between the UTR diseases and various air pollutants. Major findings are summarized below:

1. The correlation coefficients between air pollutants indicated a strong relationship between NOx and CO and moderate correlation between O_3 and CO.
2. The UTR diseases and population density were not correlated.
3. The UTR diseases and the air pollutants (NOx, SO_2, and CO) exhibited a strong association.

Suggestions for further analyses include a more refined examination of air pollutant concentrations and other factors leading to pollution, such as vehicle emissions. The monthly or daily air pollutant concentrations may be used to examine in greater detail occurrences of the UTR diseases. Vehicle emission data may be obtained from the transportation fluxes database to further the understanding of the UTR diseases and air quality.

Acknowledgements

The authors would like to thank the Environment Protection Administration, Department of Statistics, Ministry of Interior, and the Bureau of Health Promotion of the Department of Health for the provision of data for this study. Thanks also go to the National Science Council for the funding of this research (project no. NSC932320-B-434-001).

References

[1] Boezen HM, Van der Zee SC, Postman DS, et al. (1999) Effect of ambient air pollution on upper and lower respiratory symptoms and peak expiratory flow in children. Lancet 1999; 353:874-878

[2] Brunekreef B (2002) Particulate matter and lung function growth in children: a 3-yr follow-up study in Austrian schoolchildren (correspondence). Eur Respir J 2002; 20:1354-1355
[3] Edwards AL (2002) The Correlation Coefficient. In: An Introduction to Linear Regression and Correlation. San Francisco, CA: W. H. Freeman, Ch4, pp. 33-46, 1976
[4] Fisher NI, Lewis T, Embleton BJJ (1987) Statistical Analysis of Spherical Data, Cambridge University Press, pp 329
[5] Gibbons JD (1997) Nonparametric Methods for Quantitative Analysis, American Science Press, pp 537
[6] Shepard D (1968) A two-dimensional interpolation function for irregularly-spaced data, Proc. 23rd National Conference ACM, ACM, pp 517-524
[7] Woolcock AJ and Peak JK (1997) Evidence for the increase in asthma worldwide, Ciba Found Symp., p 206:122-134

Spatial Epidemiology of Asthma in Hong Kong

Franklin F.M. So[a] and P.C. Lai [b]

[a] Experian Asia Pacific, Experian Limited
[b] Department of Geography, The University of Hong Kong

Abstract: This chapter employs Geographic Information Systems (GIS) in a study of asthma, a major respiratory disease in Hong Kong. Hospital admissions records for the period 1996-2000 were obtained from the Hong Kong Hospital Authority. Residential addresses of patients were geocoded in a GIS and aggregated to make reference to the demographic statistics in census enumeration units. Disease maps were used to explore the relationship between the Standardized Morbidity Ratio (SMR) and three environmental factors of population density, land use, and air pollution. A strong positive correlation was found between asthma incidence and high-density residential areas and industrial development. A lack of direct association between asthma admissions and air pollution indicates that the relationship may not be that explicit from the spatial context.

Keywords: geographic information system, GIS, respiratory disease, asthma, spatial epidemiology, Hong Kong

1 Introduction

The prevalence of asthma[1] is increasing worldwide (WHO 2000; Wieringa et al., 2001; Corburn et al., 2006). The WHO (2000) estimated that about 100- 150 million people around the world were suffering from asthma and that the economic cost exceeded those of tuberculosis and HIV/AIDS combined. Asthma epidemiology reported by the medical and clinical profession has always emphasized individual and family-based risk factors

[1] Asthma is a disease characterized by recurrent attacks of breathlessness and wheezing. This condition is due to inflammation of the air passages in the lungs that affects the sensitivity of the nerve endings in the airways so they become easily irritated. In an attack, the lining of the passages swell causing the airways to narrow thereby reducing the flow of air in and out of the lungs (WHO, 2000).

such as indoor allergen exposures, tobacco smoking, and chemical irritants (Clark et al., 1992; Richard et al., 2000; Skoner, 2001). A number of epidemiological studies also identified air pollution as another risk factor (Schwartz, 1993; Koenig, 1999; Swift and Foster, 1999; Donaldson, et al., 2000; Samet et al., 2004).

Spatial epidemiology, which describes and identifies spatial variations in disease risks through map visualization and exploratory analysis, represents a new frontier of epidemiological research (Elliott et al., 2000; Berke, 2005; Ostfeld et al., 2005). The spatial context facilitates a close-up examination of diseases against their surrounding and neighborhood environmental exposures (Corburn et al., 2006). The approach provides a way to examine another spectrum of risk factors yet to be explored from the clinical perspective (Meade and Erickson, 2000; Jackson, 2003; Fritz and Herbarth, 2004; Kimes, et al., 2004; Oyana, et al., 2004). Advances in digital technologies have facilitated the use of Geographic Information Systems (GIS) in undertaking spatial epidemiological research (Ostfeld et al., 2005). GIS is a tool to input, store, retrieve, manipulate, analyze and output geographic or mappable data (Marble and Peuquet, 1983; DeMers, 1997). It facilitates three levels of analyses: (i) visual - to locate and plot disease cases, (ii) spatiotemporal – to integrate and link cases and environmental features in space and time, and (iii) geostatistical – to evaluate the disease pattern by means of spatial and statistical inference.

The limited number of spatial epidemiological studies in Hong Kong definitely underscores research gaps in this region. Applications of geographic concepts and techniques to health studies in Hong Kong came to exist only in the mid-1990s with the first series of medical maps presenting spatial patterns of malignant disease mortalities between 1979 and 1993 (Lloyd et al., 1996). These research studies mainly employed visual display and map presentation but with limited data manipulation and analytical capabilities (Lloyd, et al. 1996; Wong, et al., 1999). It was not until recently in the 21st century that disease patterns in Hong Kong were analyzed from the spatial perspective (So, 2002; Lai, et al., 2004).

Spatial epidemiological studies in Asian countries are comparatively fewer in number and not as methodologically advanced as those of the western world. While some studies conducted in the United States and other developed nations have suggested that urban areas contribute more to asthma, this observation is not evident in Hong Kong. It is believed that there is much room for exploring the use of GIS in spatial epidemiological studies of Hong Kong. The objectives of this chapter are twofold: (i) to visualize the spatial distribution of asthma, and (ii) to examine the spatial association of asthma occurrence against three environmental risk factors namely, population density, land use, and air pollution. Essentially, residential addresses of patients were mapped to represent disease events as points which were then studied against other geographic data, such as air pollution and land use surfaces. Likewise, the analysis was approached at the neighborhood level using the census subdivision of Tertiary Planning Units (TPUs) to present population density.

2 Methodology

Hong Kong (22°15'N, 114°10'E) bordering the South China Sea in East Asia, is a densely populated city of 6.9 million occupying a total land area of 1,098 square kilometers (Census and Statistics Department, 2000, p. 346). The population density in 2004 was 15 and 40 times more than Japan and China respectively (Central Intelligence Agency, 2005). Respiratory diseases were amongst the top-three killer diseases in Hong Kong (Hospital Authority, 2000, p. 10), with a cumulative death rate of over 90 per 100,000 population between 1996 and 1999. A sharp increase was evident in hospital admissions of respiratory illness which nearly doubled in ten years from 61,000 in 1988 to 118,000 in 1998 (Hospital Authority, 1999, p. 19).

The 1996-2000 admission records of 44 public hospitals under the management of the Hospital Authority formed the primary data source of this study. Records of private hospitals were not included because of inconsistencies in the data collection and record keeping protocols. Moreover, public hospital records accounted for over 80 percent of disease cases in Hong Kong (Hospital Authority, 2000, p. 67). Each record contains a patient's age, gender, residential address, disease code according to the International Code of Disease Version 9 (ICD-9 code), and dates of admission and discharge. The address data provide crucial information to link the home locations of patients, while age and gender enable further breakdowns of patients into subgroups. The study was also confined to patients discharged with asthma (ICD code = 493). A patient admitted repeatedly into hospitals with the same disease and within the study period of 1996-2000 was counted only in the first occurrence. The total records amounted to 52,020 admissions after having removed 754 repeat admissions of the same person (representing 1.13% of total) and discounting the 1,609 cases, or 3% of patients, with problem addresses.

The life expectancy of females in Hong Kong was 84.6 compared to 78.4 for males (Hospital Authority, 2003). It was found necessary to standardize the crude rates by gender and age groups to account for age-sex difference in disease impacts. The Standardized Morbidity Ratio (SMR in

Eqn. 1) was employed to compute ratios of the at-risk population in an area by comparing the observed number of patients against the expected counts of various age groups by gender.

$$SMR = \frac{N}{\sum R_{si} \times P_i} \qquad \text{Eqn. (1)}$$

where N = total number of patients in the observed population
R_{si} = age-gender specific morbidity rate of age-gender interval i in the standard population
P_i = the population of age-gender interval i in the observed population

The SMR measure expresses the difference between morbidity experience of the population under study and the experience of that population as it would be if it experienced the age-specific rates of the comparison population. A ratio of greater than 1.0 indicates greater than expected morbidity has occurred while less than 1.0 indicates lower than expected morbidity. The decimal fraction shows the percentage comparison. An SMR of 1.12, for example, infers 12% more than the expected number of patients has occurred.

GIS was employed to geocode the residential addresses by assigning data items to some locations or geographic positions in Earth space (Croner et al., 1996). However, address entries must first be standardized to eliminate inconsistencies (such as mis-spellings or incomplete records) since the patients were not provided with guidelines or rules in entering their addresses. It was possible to geocode about 89.3 percent of the raw records without much difficulty and painstakingly geocode 7.5 percent of the records by interactive means, which required substantial manual inputs. The remaining 3 percent of addresses could not be geocoded either because of missing details or incorrect entries of non-existing streets or buildings.

Maps were employed to visualize spatial patterns of asthma using the ArcGIS 9.1 software. Geocoded point data derived from address matching formed the essential data for further analysis. Three levels of mapping and spatial analyses were undertaken: point, Tertiary Planning Unit[2] (TPU), and District[3]. While point data can reveal raw distributional patterns, com-

[2] Tertiary Planning Unit (TPU) is a census enumeration unit demarcated by the Planning Department for the territory of Hong Kong. Each TPU is sub-divided into a number of Street Blocks (SB) for urban areas and Village Clusters (VC) for rural areas.

[3] There are at present 18 districts in Hong Kong, 9 in the urban areas and 9 in the New Territories, including Central and Western, Eastern, Islands, Kowloon City, Kwai Tsing, Kwun Tong, Northern, Sai Kung, Sham Shui Po, Shatin,

parison of these data with demographic statistics must be made in reference to census enumeration units by TPUs and Districts. The point data were therefore aggregated by areal units using the point-in-polygon operation in a GIS to enable the making of thematic maps (Gatrell, 2002, p. 51). This study employed the 2001 population census data of 139 TPUs as defined by the Census and Statistics Department for finer comparison and 18 districts as defined by the District Council for territorial-wide analysis.

GIS was employed to facilitate the point-in-polygon and polygon-on-polygon overlay operations. The geocoded point data were overlaid in a GIS to examine variation of the SMR of asthma by the 18 Districts and 139 TPUs of Hong Kong. Analysis at the District level provides a general overview, and more detailed spatial variations are reflected at the TPU level. It was postulated that asthma occurrences are associated with population density and land use types. The polygon-on-polygon overlay operation was used to examine the association of SMR of each district and TPU against the population density as well as the land use patterns defined by the Planning Department.

The association between asthma and air pollution was analyzed. Air pollution data of Hong Kong were derived from a total of 14 monitoring sites that measure, on a daily basis in Hong Kong since 1999, concentrations of major pollutants, including the ambient respirable suspended particulate, sulphur dioxide, carbon monoxide, ozone and nitrogen dioxide (Environmental Protection Department, 2002). The Air Pollution Index (API) summarizes the severity of the pollution level and is calculated by comparing the measured concentrations of major air pollutants against their respective health related Air Quality Objectives established under the Air Pollution Control Ordinance. The mean seasonal API values were calculated for the 14 sites and the Inverse Distance Weighting (IDW) method (Chung and Yun, 2004) used to simulate a grid-based air pollution surface for each of the four seasons. Monthly hospital admission rates of asthma patients for each district were regrouped to reflect seasonal changes and these were compared against the overall average (i.e. annual average admission rate). Seasonal variations of asthma admissions of each district were reduced to percentages above or below the overall average. These seasonal figures were mapped against the API surfaces for the respective seasons and their correlation coefficients computed.

Statistical methods were used to account for association between asthma and three environmental variables of population density, types of land use, and air pollution. The distribution of asthma was assumed to have a Pois-

---

Southern, Tai Po, Tsuen Wan, Tuen Mun, Wan Chai, Wong Tai Sin, Yau Tsim Mong, and Yuen Long.

son probability distribution because the number of occurrences in two mutually exclusive intervals is independent. The probability of asthma occurring within a small interval is small and proportional to the length of the interval, i.e. the event is rare. In addition, the measure to account for data correlation in this research is the Pearson's product moment correlation coefficient.

3 Results

The SMRs of asthma patients for the 18 districts of Hong Kong were mapped in Figure 1 in which areas of higher morbidity have darker tones and those of lower morbidity have lighter tones. The mean SMR of 18 districts was 0.96 with a standard deviation of 0.19 at 99 percent confidence interval. Figure 1 revealed that Kwai Tsing, the central part of Hong Kong, had 36 percent higher than the expected number of asthma patients. Western Hong Kong (including Yuen Long, Tuen Mun, and Lantau) had lower than the expected number of asthma patients.

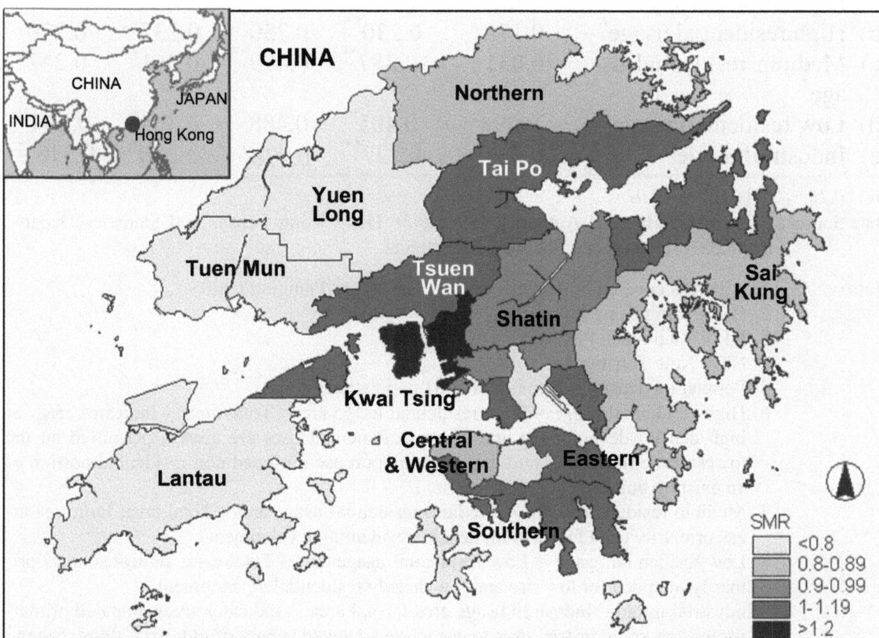

Fig. 1. Distribution of asthma patients according to standardized morbidity ratios (SMR) by Districts of Hong Kong

SMR values were correlated against population density at the district and TPU levels (Table 1). The district level analysis indicated a positive but marginally insignificant (close to 90 percent confidence level) correlation between disease morbidity and population density (r=0.329, p=0.103). Given that the maximum plot ratio (i.e. Gross Floor Area / Population Number) varies for different land use types, SMRs were further correlated with densities of different land uses (classified as high-, medium- and low-residential, and industrial in accordance with the statutory land use planning put forth by the Planning Department). Although not statistically significant, it was interesting to note that SMR correlated positively with high-density residential development (r=0.178) and negatively with low-density residential development (r=-0.094). The above results warranted more detailed examination at the TPU level.

Table 1. Correlation coefficients of SMR against population density and land use types

	Districts	Tertiary Planning Units			
	All[1] (N=18)	All[1] (N=139)	HK[2] (N= 37)	KLN[3] (N= 44)	NT[4] (N= 58)
(a) Population density[5]	0.329^{+}	0.174^{*}	0.149^{+}	0.127	0.262^{*}
(b) High residential usage[6]	0.178	0.230^{**}	0.230^{+}	0.237^{+}	0.270
(c) Medium residential usage[7]	-0.031^{+}	-0.387^{**}	-0.499^{**}	-0.351^{**}	-0.242^{+}
(d) Low residential usage[8]	-0.094	-0.405^{**}	-0.488^{**}	-0.364^{**}	-0.434^{**}
(e) Industrial usage[9]	0.064	0.227^{**}	0.278^{+}	0.208^{*}	0.318^{**}

$^{+}p < 0.10$, $^{*}p<0.05$, $^{**}p<0.01$

Data Source: Hong Kong Hospital Authority 1996-2000; Hong Kong Census and Statistics Department 2001; Hong Kong Planning Department.

Note:
1 All areas covered by 18 Districts or 139 Tertiary Planning Units
2 HK = Hong Kong Island
3 KLN = Kowloon Peninsula
4 NT = New Territories
5 Population density = Number of 2001 population / Area
6 High residential usage = High residential usage area / Total area; Indicates areas of high density development and where commercial uses are always permitted on the lowest three floors of a building or in the purpose-designed non-residential portion of an existing building
7 Medium residential usage = Medium residential usage area / Total area; Indicates areas primarily used for medium density residential development
8 Low residential usage = Low residential usage area / Total area; Indicates areas primarily intended for low rise and low density residential development
9 Industrial usage = Industrial usage area / Total area; Indicates areas intended primarily for general industrial uses to ensure an adequate supply of industrial floor space to meet demand from

Mapping of SMRs and correlation analyses at the TPU level were much more revealing than the generalized district-based statistics. Figure 2 showed much variation in the SMR readings by TPUs in each of the 18 districts. The mean SMR value of 139 TPUs was 0.92, with a standard deviation of 0.32 at 95 percent confidence interval. It was also interesting to note that while the mean SMR of three geographic constituencies (2a - Hong Kong Island, 2b - Kowloon Peninsula, and 2c - New Territories) were similar, SMR values of TPUs in Hong Kong Island varied much more (standard deviation of 0.4) than for the Kowloon Peninsula and New Territories (with standard deviations of 0.28 and 0.29 respectively). Besides Kwai Tsing which registered a high SMR (at both district and TPU levels), there were actually a few TPUs with high SMRs in Hong Kong Island and the Kowloon Peninsula. Details rendered through TPUs (Figure 2) seem superior to those of Districts (Figure 1) in revealing the spatial variation of disease morbidity.

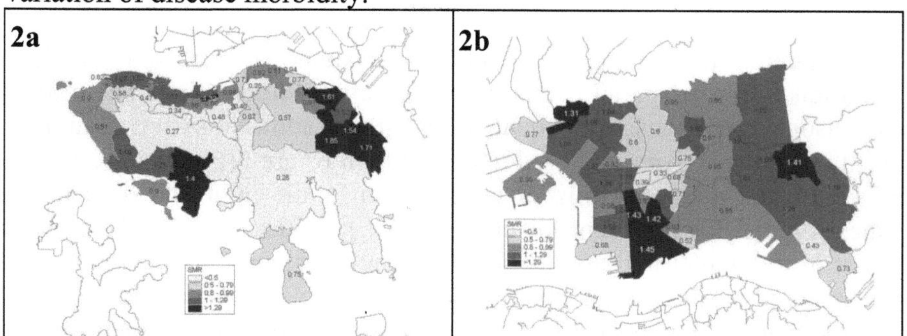

N	Mean	Standard Deviation	Confidence Interval	N	Mean	Standard Deviation	Confidence Interval
37	0.90	0.40	± 0.13	44	0.97	0.28	± 0.08

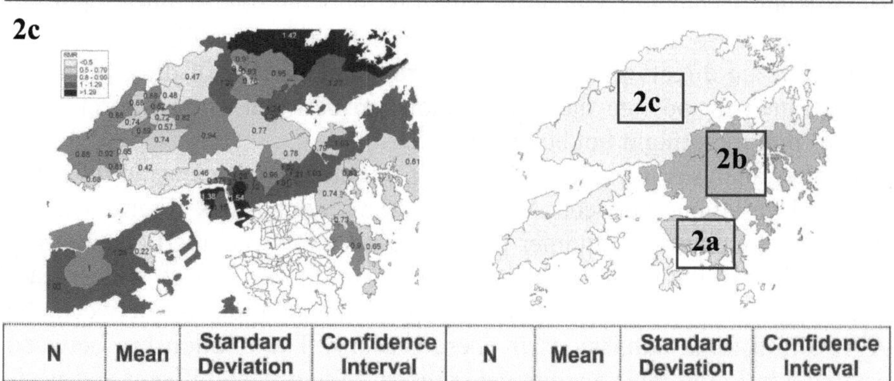

N	Mean	Standard Deviation	Confidence Interval	N	Mean	Standard Deviation	Confidence Interval
58	0.91	0.29	± 0.08	139	0.92	0.32	± 0.05

Note: 3a = Hong Kong Island; 3b = Kowloon Peninsula; 3c = New Territories
Data Source: Hospital Authority 1996-2000; Census and Statistics 2001.

Fig. 2. Distribution of asthma patients according to standardized morbidity ratios (SMR) by Tertiary Planning Units of Hong Kong

Table 1 presents correlation values broken down by three geographic constituencies in comparison to the whole of Hong Kong. There is a clear indication of a statistically significant correlation between SMR and population density in Hong Kong (r=0.174, p<0.05). Further analyses broken down by land use types indicated positive correlation with high-density residential and industrial development (r=0.23, p<0.01 and r=0.227, p<0.01) and negative correlation with medium- and low- density residential development (r=-0.387, p<0.01 and r= -0.405, p <0.01), all of which were statistically significant at the 99 percent confidence interval. Interestingly, the geographic constituencies contributed in varying degrees to the highly significant relationships in each land use types. In particular, the negative correlation between SMR or asthma morbidity and low-density residential development was significant in all three geographic constituencies.

The attempt to examine the possible association between asthma and air pollution from the spatial perspective was rather illuminating. The effects of seasonal variations of air quality on asthma admissions were summarized in Figure 3. Areas colored in dark brown or grey (Figures 3a, 3c and 3d) indicated the worsening of air pollution while light and dark green tones (Figure 3b) represent lower APIs. Air pollution in Hong Kong was worst in the winter and acceptable in the summer months. The maps also underscored northern Wan Chai as the pollution black spot in Hong Kong. At first glance, Figure 3d revealed that adverse atmospheric conditions in the winter had a parallel effect on asthma admissions. However, the highest APIs were recorded in Eastern, Central and Western districts of the Hong Kong Island and Yau Tsim Mong in the Kowloon Peninsula but the greatest increase in admissions was not reflected in these districts (as evident from the directions and sizes of arrows). This lack of direct association seemed to indicate that the relationship between asthma admissions and air pollution might not be explicit.

Further analyses using the Pearson's correlation coefficients highlighted seasonal variations between APIs and asthma admissions (Figure 4). Although the winter and summer months respectively recorded consistently high or low pollution levels, the range of values was not as widespread. Statistical results did not indicate a significant relationship between the APIs and asthma admissions in these seasons. The relationship between asthma admission rates against generalized APIs was significant and positive for spring (r=0.534, p<0.01) but negative for autumn (r=-0.581, p<0.01). It is suspected that unstable air currents brought about by the monsoons in autumn contributed to this lack of association. The overall plot however clearly indicates a positive relationship with statistical significance (r=0.652, p<0.01).

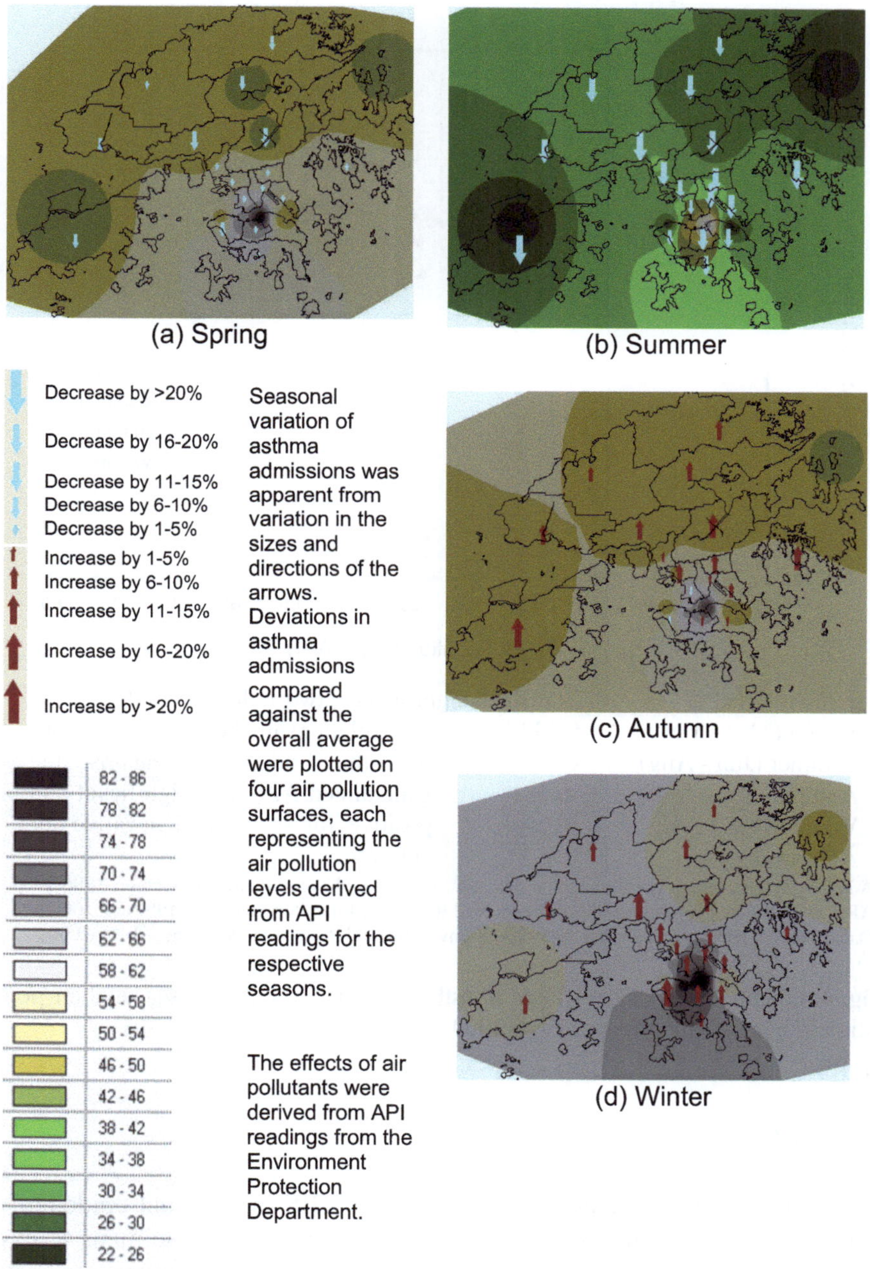

Fig. 3. Effects of seasonal variations of air quality on asthma admissions in Hong Kong, 1996-2000

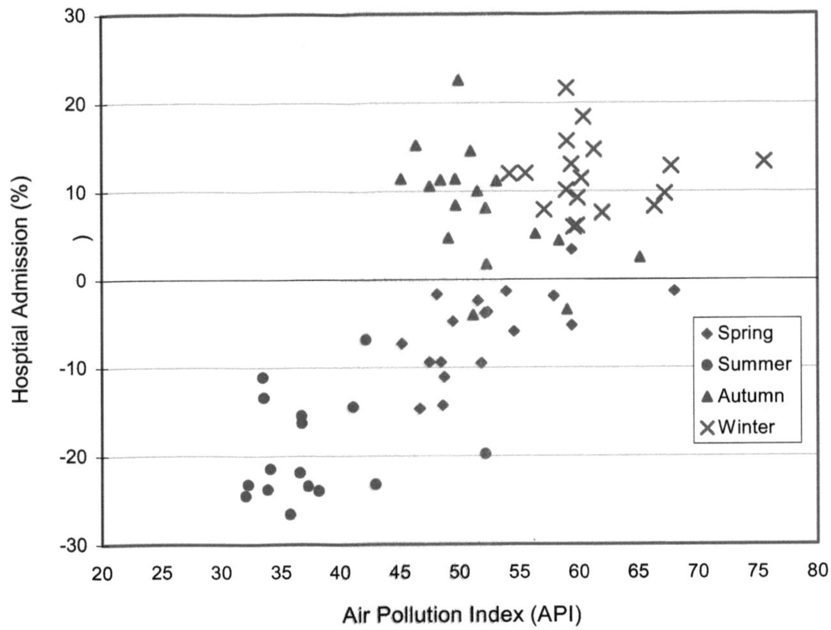

Time Interval	Statistical significance, two-tailed: *p<0.01	
Spring (Mar - May)	0.534*	The relationship between deviations in asthma admissions against the generalized APIs was significant and positive for spring and negative for autumn.
Summer (Jun - Aug)	0.274	
Autumn (Sep - Nov)	-0.581*	
Winter (Dec - Feb)	-0.296	
All seasons	0.652*	

Admission: Percent of hospital admissions of asthma cases above or below the average, 1996-2000.
API: Generalized Air Pollution Indices sourced from the 14 monitoring sites in 1999 and 2000.
Data Source: Hospital Authority 1996-2000; Environmental Protection Department 1999-2000.

Fig. 4. Graphic plots of deviations in asthma admissions against seasonal air pollution indices of Hong Kong, 1996-2000

4 Discussion

This research has demonstrated the use of GIS in a spatial epidemiological study of asthma in Hong Kong making use of disease maps to examine the causative relationship between asthma and environmental risk factors, i.e. population density, land use, and air pollution. This study is the first attempt in Hong Kong to use hospital morbidity data with patient addresses, particularly down to individual building blocks, to geocode and analyze patterns of diseases in a GIS. While the approach of point mapping that displays residential locations of patients has raised concerns regarding privacy and confidentiality of health records (Rushton, 1998), the availability of such data enables nested case-control or case-cohort study to be conducted for direct study within a large-scale population based cohort (Diggle, 2003; Elliott and Wartenberg, 2004). Point mapping at the initial stage was adopted to compile area-based statistics by district and TPU. It was shown that data summarized by 139 TPUs provided sufficient detail to characterize spatial variation of disease morbidity.

The use of exact address location as the data unit may also lead to the problem of epidemiological inference (Lawson, 2001) because a case address used to represent residential exposure to environmental risk may not be comprehensive enough to reflect the individual exposure occurring at work, at school, or at regular weekend visits (Wakefield, et al., 2001). Given that technologies can handle very detailed accounts of human activities, the question of how far 'society' wants to go in the use of personal data to increase our understanding of disease, and exposure, must be posed. Considering all of the above, mapping by patient addresses is an acceptable and practical option.

The flexibility to partition data sets by various levels of detail by automated means in a GIS has greatly extended our ability to probe spatial data. Users of GIS, however, must take time to examine the input data and make necessary adjustments to the data before further analysis. Various cartographic and geo-statistical techniques were employed to enhance understanding of the spatial event. Some cartographic presentations shown in this research are experimental in nature as they have not been attempted by other researchers. The air pollution surfaces were interpolated and extrapolated from APIs recorded from 14 monitoring stations. Despite extensive epidemiological evidence of the adverse effect of air pollution on asthma, this study did not reveal a direct association, perhaps due to the sparseness of the API data points. Although mobile air pollution monitoring platforms have been developed to supplement the fixed number of monitoring sites (Department of Chemical Engineering HKUST, 2002), this study did not have access to these data. Other contributing factors of air pollution (such as terrain, wind, and traffic) were not accounted for in this study. The relationship between these risk factors can also be further examined in the three-dimensional (3D) modeling environment of a GIS. For instance, the relationship between respiratory diseases and the ventilation of air conditioning systems in offices or buildings can be investigated in 3D. This method would be particularly valuable in Hong Kong because of its hilly terrain and the abundance of high-rise buildings. The topographic effect of air pollution, wind direction, and wind speed on the distribution of disease patients should also be investigated in the future.

The feasibility of the approach was largely dependent on the data quality. It should be noted that there were quite a number of data problems in conducting this study. Attempts were made to resolve these issues in a satisfactory manner, but some of the problems would never have arisen had certain data been included in the process of data collection. First, given that our study employed primarily secondary data sources, some issues on data availability and comparability were unavoidable. In addition, patient data of this research originated from admission records of public hospitals under the management of the Hospital Authority. Similar data of private hospitals, or clinics other than hospitals, were not included in this study because of data inconsistency and resource constraints. It is arguable that the incomplete coverage of all of the hospitals and clinics may result in the data set not being sufficient to reflect the true picture of disease patterns. Patient data might need to be derived from medical records of clinics other than hospitals if the pattern of morbidity was to be better represented. Second, the use of a mixture of data (patient data 1996-2000, census data 2001, and land use data 2001) also raises the concerns of data representativeness and the extent to which epidemiological interpretation and inference can be made. The morbidity of the respiratory diseases could be studied by examining changes in spatial patterns of different time intervals, were longer periods of medical records available. This research considered only five years of data from the Hospital Authority. Finally, socioeconomic variables (e.g. income, occupation, educational level, etc.) were not taken into account. Areas with a high population density tend to imply less than desirable living conditions which may better explain the relationship between asthma morbidity and poverty households. This aspect was examined in an earlier study (So, 2002) which we shall address further in another paper.

5 Conclusion

This study demonstrated the use of three levels of spatial epidemiological approaches to examining asthma occurrences. Firstly, the visual analysis helps locate and plot disease cases. Visual presentation is made possible with geocoding of patient addresses which facilitates rapid assessment of disease clusters. Secondly, the spatiotemporal approach is highlighted by a space-time examination of disease spread against seasonal air pollution patterns. While the results did not reveal a direct association between asthma and pollution, the point-in-polygon and polygon-on-polygon overlay methods are instrumental in uncovering and perhaps alerting possible association for further scrutiny. Finally, the geostatistical method was applied to assess the disease pattern against land use types by means of spatial and statistical inference. The approach allows for the quantification of possible spatial association, besides visual evidence, which offers an objective basis for comparative analyses.

Acknowledgment

We are indebted to the Hui Oi Chow Trust Fund of the University of Hong Kong for the generous support. Data sources for the research include the Hong Kong Hospital Authority, Hong Kong Census and Statistics Department, Hong Kong Planning Department, and Hong Kong Environmental Protection Department.

References

[1] Berke O (2005) Exploratory spatial relative risk mapping, Preventive Veterinary Medicine 71, 173-182
[2] Bowling A (1997) Research Methods in Health: Investigating Health and Health Services. Philadelphia: Open University Press
[3] Chau CK, Tu EY, Chan DWT, Burnett, J (2002) Estimating the total exposure to air pollutants for different age groups in Hong Kong, Environmental International 27, 617-630
[4] Chung U and Yun JI (2004) Solar irradiance-corrected spatial interpolation of hourly temperature in complex terrain, Agricultural and Forest Meteorology 126, 129–139
[5] Census and Statistics Department (2000) Hong Kong Annual Digest of Statistics. Hong Kong: Government Printer
[6] Census and Statistics Department (2001) Hong Kong 2001 Population Census TAB on CD-ROM. Census and Statistics Department, HKSAR Government
[7] Central Intelligence Agency (2005) World Fact Book, Available at http://www.cia.gov/cia/publications/factbook/index.html [Accessed on 23 February 2005]
[8] Clark TJH, Godfrey S, Lee, TH (1992) Asthma. London: Chapman and Hall Medical, 3rd Edition
[9] Corburn J, Osleeb J, Porter M (2006) Urban asthma and the neighborhood environment in New York City, Health and Place 12, 167-179

[10]Croner CM, Sperling J, Bromme FR (1996) Geographic Information System (GIS): New Perspectives in Understanding Human Health and Environmental Relationships, Statistics in Medicine 15, 1961-1977
[11]DeMers MN (1997) Fundamentals of Geographic Information Systems. New York: Wiley
[12]Department of Chemical Engineering, Hong Kong University of Science and Technology (2002) Mobile, real-time air monitoring platform, Available at http://www.ust.hk/~webiesd/Project%20Frame.htm [Accessed on 23 February 2005]
[13]Diggle PJ (2003) Statistical analysis of spatial point patterns, London: Arnold; New York: Distributed by Oxford University Press, Chapter 4, 42-62
[14]Donaldson K, Gilmour IM, MacNee W (2000) Asthma and PM10, Respiratory Research 1, 12-15
[15]Environmental Protection Department, 2002. API and Air quality, Available at http://www.info.gov.hk/epd/english/environmentinhk/air/air_quality/air_qua lity.html [Accessed on 23 February 2005]
[16]Elliot P, Wakefield JC, Best NG, Briggs DJ (2000) Spatial Epidemiology: Methods and Applications. Oxford, UK: Oxford University Press
[17]Elliot P and Wartenberg D (2004) Spatial Epidemiology: Current Approaches and Future Challenges, Environmental Health Perspectives 12 (9), 998-1006
[18]Fritz GJ and Herbarth O (2004) Asthma disease among urban preschoolers: an observational study, International Journal of Hygiene and Environmental Health 207, 23-30
[19]Gatrell AC (2002) Geographies of Health: An Introduction. Oxford, UK: Blackwell Publishers Ltd
[20]Hospital Authority (2003) Hospital Authority Statistical Report 2002-2003. Hong Kong: Hospital Authority
[21]Hospital Authority (2000) Hospital Authority Statistical Report 1999-2000. Hong Kong: Hospital Authority
[22]Hospital Authority (1999) Hospital Authority Statistical Report 1998-1999. Hong Kong: Hospital Authority
[23]Jackson LE (2003) The relationship of urban design to human health and condition, Landscape and Urban Planning 64, 191-200
[24]Kimes D, Ullah A, Levine E, Nelson R, Timmins S, Weiss S, Bollinger ME, Blaisdell C (2004) Relationships between pediatric asthma and socioeconomic/urban variables in Baltimore, Maryland, Health and Place 10(2), 141-152
[25]Lai PC, Wong CM, Hedley AJ, Lo SV, Leung PY, Kong J, Leung GM (2004) Understanding the Spatial Clustering of Severe Acute Respiratory Syndrome (SARS) in Hong Kong, Environmental Health Perspectives 112(15), 1550-1556
[26]Lawson AB (2001) Statistical Methods in Spatial Epidemiology. Chichester: John Wiley and Sons Ltd
[27]Lloyd OL, Wong TW, Wong SL, Yu TS (1996) Atlas of Disease Mortalities in Hong Kong for the Three Five-year Periods in 1979-93. Hong Kong: Chinese University Press

[28]Marble DF and Peuquet DJ (1990) Introductory readings in geographic information systems. London: Taylor and Francis
[29]Meade MS and Earickson RJ (2000) Medical geography. London: Guildford Press
[30]Ostfeld RS, Glass GE, Keesing F (2005) Spatial epidemiology: an emerging (or re-emerging) discipline, Trends in Ecology and Evolution 20(6), 328-336
[31]Oyana TJ, Rogerson P, Lwebuga-Mukasa JS (2004) Geographic clustering of adult asthma hospitalization and residential exposure to pollution sites in Buffalo neighborhoods at a U.S.-Canada Border Crossing Point in American Journal of Public Health 94(7), 1250-1257
[32]Richard C and Beasley W, Pearce NE (2000) Epidemiology of Asthma Mortality. In: Giembycz MA, O'Connor BJ (Eds) Asthma: Epidemiology, Anti-Inflammatory Therapy and Future Trends, Basel: Birkhäuser Verlag, Chapter 1, 1-24
[33]Rushton G (1998) Improving the Geographical Basis of Health Surveillance using GIS. In: Gatrell, A.G., Loytonen, M. (Eds) GIS and Health. London; Philadelphia: Taylor and Francis, Chapter 5, 63-80
[34]Samet JM, Dominici F, Curriero FC, Coursac I, Zeger SL (2000) Fine particulate air pollution and mortality in 20 US cities, 1987-1994, N England J Med. 343, 1742-1749
[35]Schwartz J (1993) Particulate air pollution and chronic respiratory disease in Environ Res 62, 7-13
[36]Skoner DP (2001) Asthma and Respiratory Infections. New York; Basel: Marcel Dekker
[37]Swift DL and Foster WM (1999) Air Pollutants and the Respiratory Tract. New York: Marcel Dekker
[38]So FM (2002) An Application of Geographic Information Systems in the Study of Spatial Epidemiology of Respiratory Diseases in Hong Kong, 1996–2000. MPhil Thesis. Hong Kong: The University of Hong Kong
[39]Wakefield SEL, Elliott SJ, Cole DC, Eyles JD (2001) Environmental risk and (re)action: air quality, health, and civic involvement in an urban industrial neighborhood, Health and Place 7(3), 163-177
[40]Wieringa MH, Verimerie PA, Van Bever HP, Helen VJ, Weyler JJ (2001) High occurrence of asthma-related symptoms in an urban than a suburban area in adults, but not in children in Eur Respir Journal 17, 422-427
[41]Wong TW, Lau TS, Yu TS, Neller A, Wong SL, Tam W, Pang SW (1999) Air pollution and hospital admissions for respiratory and cardiovascular diseases in Hong Kong, Occupational and Environmental Medicine 56(10), 679-683
[42]World Health Organization (2000) Factsheets of bronchial asthma, Available at http://www.who.int/mediacentre/factsheets/fs206/en/ [Accessed on 23 February 2005]

Disease Modeling

An Alert System for Informing Environmental Risk of Dengue Infections

Ngai Sze Wong [a], Chi Yan Law [a], Man Kwan Lee [a], Shui Shan Lee [c] and Hui Lin [a,b]

[a] Geography & Resource Management Department, Chinese University of Hong Kong, Hong Kong, China
[b] Institute of Space and Earth Information Science, Chinese University of Hong Kong, Hong Kong, China
[c] Centre for Emerging Infectious Disease, Chinese University of Hong Kong, Hong Kong, China

---

Abstract: Dengue is a mosquito-borne infection the incidence of which varies with the environment and climate. Knowingly, the reported incidence of dengue is an insensitive indicator of infection risk in a locality. The Ovitrap Index has been in use in many countries. This index is a measurement of mosquito eggs in specified geographic locations which, in turn, reflects the distribution of *Aedine* mosquitoes, the vector for dengue. An alert system, founded on the Geographic Information System (GIS) technology, was created from a synthesis of geospatial data on ovitrap indices in Hong Kong. The inter-relationship between ovitrap indices and temperature was studied. This inference forms the rationale for the generation of weighted overlays to define risk levels. The weighting can be adjusted to set the sensitivity of the alert system. The system is operational at two levels: (1) for the general public to assist in the evaluation of dengue risk in the community, and (2) for professionals and academia in undertaking technical analysis. The alert system offers one objective means to defining the risk of dengue in a society, which would not be affected by the incidence of the infection itself.

Keywords: health, GIS, Dengue Alert System, Ovitrap Index, weighted overlay

---

1. Introduction

Dengue is a mosquito-borne infection the incidence of which varies with climate and environment. The Ovitrap Index is a measurement of mosquito eggs in specified geographic locations to reflect the distribution of the *Aedine* mosquitoes. The main objective of this chapter is to discuss an alert system built on a geographic information system (GIS) to describe the risk level of dengue infection in districts of Hong Kong.

Aedes albopictus is the local vector of dengue fever. It is a day biter (the most active times are within 2 hours after dawn and before sunset) and a weak flier (with a flight range of about 100 meters). The presence of adult Aedine mosquitoes in selected areas is detected by the oviposition trap (ovitrap), which is a simple device made of a black plastic container with a brownish oviposition paddle placed diagonally (Figure 1) (FEHD 2006).

Fig. 1. An ovitrap

A total of 38 locations in Hong Kong have been installed with ovitraps for the vector surveillance. The Ovitrap Index is measured monthly and formulated as follows (FEHD, 2006):

$$OvitrapIndex = \frac{no.ofAedes - positiveOvitrpas}{no.ofOvitraps} X100\% \qquad \text{Eqn. (1)}$$

Two types of Ovitrap Indices are available: the Area Ovitrap Index (AOI) and the Monthly Ovitrap Index (MOI). The Ovitrap Index in this chapter refers to AOI which measures the extensiveness of the distribution of *Aedine* mosquitoes in a particular area surveyed (FEHD, 2006).

There are specific preventive and control measures recommended for each of the four alert levels (Table 1):

Table 1. Level of Action to be taken (FEHD 2006)

Level	Ovitrap Index	Action to be taken
1	<5%	Closely monitor the hygienic condition to prevent breeding Conduct weekly inspection to identify breeding/ potential breeding places and eliminate
2	5%≤O.I<20%	Public are advised to check and eliminate possible breeding places within their premises at least once a week
3	20%≤O.I<40%	Conduct special operations in addition to the regular weekly program to eliminate all breeding/potential breeding places

4	O.I≥40%	Private pest control contractor might be employed to control the mosquito problem. Other control measures by using larvicides or adulticides might be feasible

2. System Design and Development

2.1. Assumption

The alert system is designed to predict the risk of dengue infection. It is assumed that temperature and the Ovitrap Index are two factors strongly associated with the exposure to dengue fever infection and they are of great geographic significance.

2.2. Data Source

Data on Ovitrap Index for 2004 and 2005 are available from the Food and Environmental Hygiene Department (FEHD), and the 2004 and 2005 monthly mean temperatures from the Hong Kong Observatory. The base map of Hong Kong is obtained from the Geography and Resource Management Department of the Chinese University of Hong Kong.

2.3. Conceptual Design

The alert system is divided into two modules: (1) technical analysis and (2) the alert system proper. The system is designed for four levels of users. The first level is the general public with limited data access. The second level is for researchers with more data access and who may not be familiar with the operation of GIS software. Less demanding user interfaces are designed for users of the first two levels to allow for limited usage of GIS tools without difficulty. The third level is meant for researchers familiar with the operation of GIS software and who are able to conduct in-depth analyses with GIS tools. The fourth level users are those maintaining the system.

First level users can only have access to the current Ovitrap Index, risk level data and the corresponding measures by selecting an area of concern. For second level users who have little knowledge of GIS software usage, interfaces are designed to facilitate their access to additional data such as Ovitrap Indices from 2004 onwards and the buffer areas. Third level users are those familiar with GIS operations. They have easy access to and can reach more data for analyses. Data accessible by both levels two and three users include past and current statistics on the Ovitrap Index. A login system is set up to distinguish the type of users and to provide appropriate guidance on different GIS tools for undertaking analyses. For instance, the buffer function can be used to identify affected and potentially affected areas. Users at the highest level include authorized personnel for data maintenance having the rights to access, change and update the database. User interfaces are also designed to simplify their work. They can either enter the data directly to the attribute table through the interface or by loading the table from other sources.

The alert levels used in the dengue alert system are built according to guidelines of the FEHD as listed in Table 1. There is only one factor, namely the Ovitrap Index, affecting the risk level. However, results of the weighted overlay analysis offer another perspective for assessing risk levels and are recommended here to researchers for investigating the feasibility and standardization of a new set of risk levels.

2.4. Function and System Requirement

The functions envisioned for the system include the following: (i) graduated color mapping for spatial distribution analysis, (ii) buffering for showing affected and potentially affected areas, (iii) zoom-in views for close-up examination, (iv) raster calculator for raster data overlay, and (v) weighted overlay for controlling the sensitivity of the alert system. The Visual Basic Editor was used to program the user interface.

The hardware needed for the system includes a computer. The software requirement comprises the ArcMap GIS software developed by ESRI which enables users to undertake spatial data building, modeling, analysis, and map display.

2.5. System Framework

The system framework (Figure 2) shows the four user levels and three user interfaces offering an easy access to data and system functions. Level 1 users will employ the "Public" interface to learn about the alert levels and actions under each. Users of levels 2, 3 and 4 may also use functions designed for level 1 users as no login is required. Users of levels 2 and 3 can identify affected areas through the "Technical" interface. In addition, level 4 users have the right to change contents of data residing in the system through the "Data Maintenance" interface. These interfaces guide their respective users to selective functions, to databases of the Ovitrap Index and temperatures, as well as a base map of Hong Kong in different spatial extent.

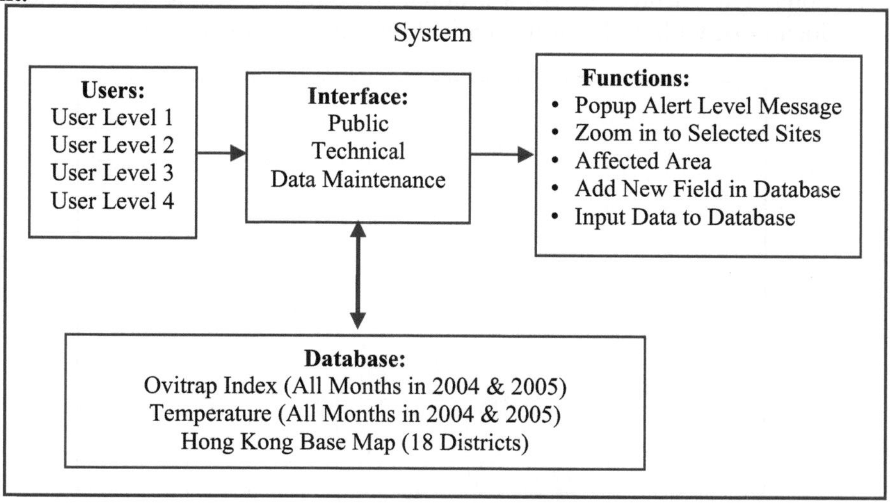

Fig. 2. System framework

2.6. System Development

The project objectives were first determined and user requirements analyzed. Schematic flow charts were drawn and modified to obtain a blueprint for the system. Upon assembling and processing the input data, GIS functions were explored. The alert system was then tested for its operational efficiency and necessary modifications made. Finally, data maintenance procedures and working details were defined.

3. Applications of the Dengue Alert System

3.1. Feature Description

Raw data on monthly temperatures for 2004 and 2005 as well as locational coordinates of sampling stations are stored in the dbf file format. The station data are geo-referenced by a GIS according to the X and Y coordinates. A new raster layer or temperature surface (Figure 3) is then created by the inverse distance weighted function provided by the ArcGIS Spatial Analyst extension. The temperature surface is estimated from monthly temperature readings at a set of disperse points representing the sampling stations. The surface provides a quick overview of temperature variation by locations, with darker tones indicating higher temperatures.

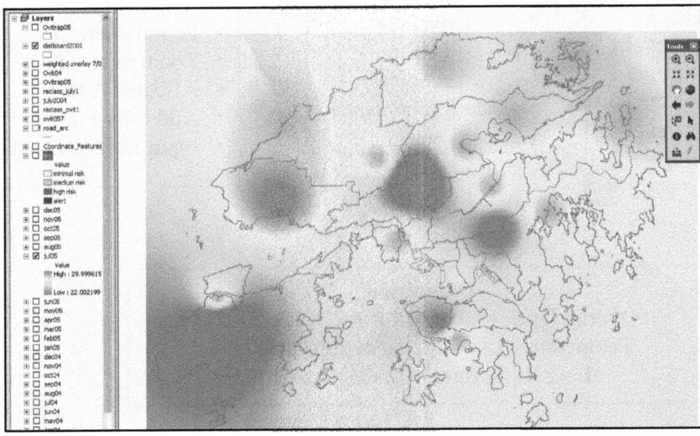

Fig. 3. A raster layer of temperature

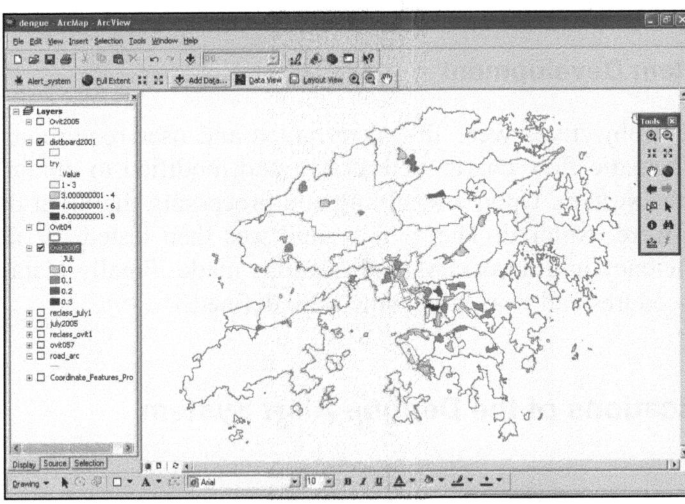

Fig. 4. A polygon layer of Ovitrap Index sites

Raw data on monthly readings of the Ovitrap Index for 2004 and 2005 are also stored in the dbf file format. These data can link with the attribute table of the geo-referenced Ovitrap Index sites shown as a polygon layer (Figure 4).

3.2. Spatial Query

The Visual Basic Editor was used to design the forms for the user interface, stored in the frm file format. The system will first require its users to indicate their "Public" or "Academic" status. Public users, upon gaining entry, can view the risk of Dengue for the whole territory of Hong Kong (Figure 5) or select a district in Hong Kong to examine in greater detail (Figure 6).

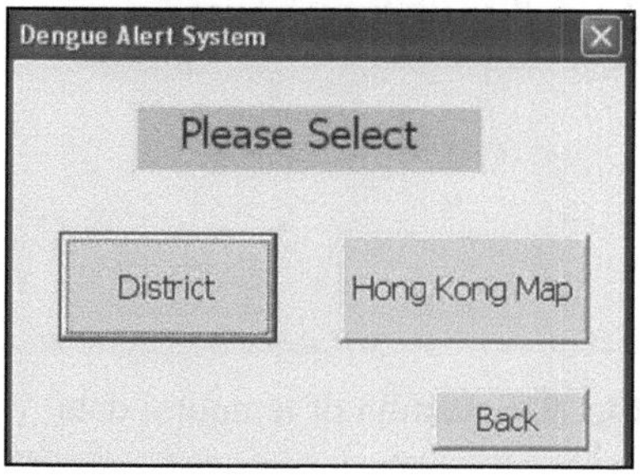

Fig. 5. A function selection box

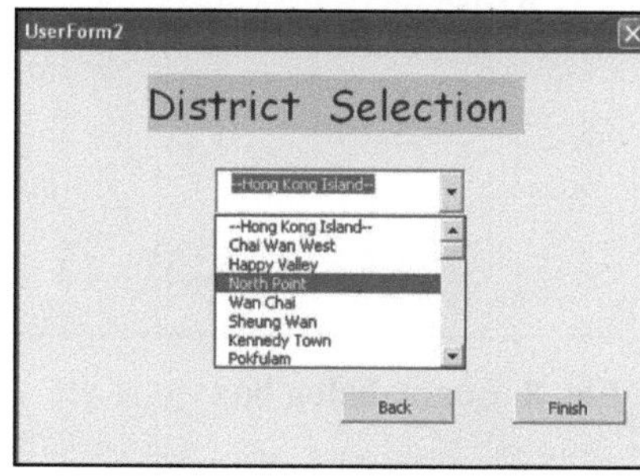

Fig. 6. Selection with a combo box

At the selection of the district option, a logical operation will be carried out and the result displayed in a dialog box to indicate the Dengue alert level, ranging from no potential danger (Level 0) to alert level (Level 4). Using Fanling as an example, the alert level shown in Figure 7 is a Level 3 and its associated advice to preventing dengue infection is also provided in the message box. Users can choose to exit the system by clicking the "Finish" button or return to select another operation by clicking the "Back" button.

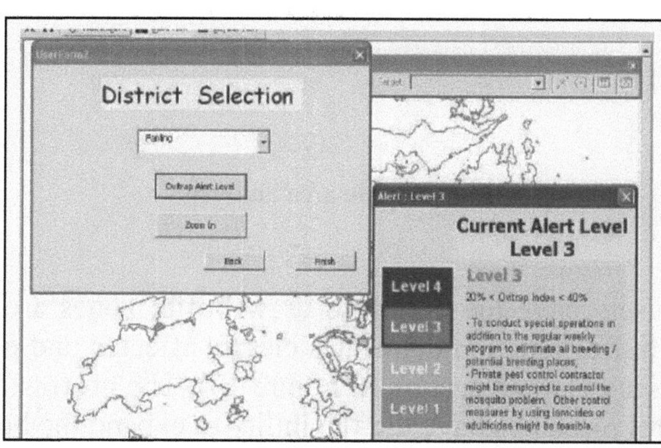

Fig. 7. Dengue alert message by district

Technical users selecting the "Academic" status are required to login with a user ID and a password (Figure 8). This login mechanism serves to protect the security of data under research. Upon gaining entry to the system, users are asked to make a selection of the year of study (Figure 9). Both contents of the database and selections are expected to grow in the future.

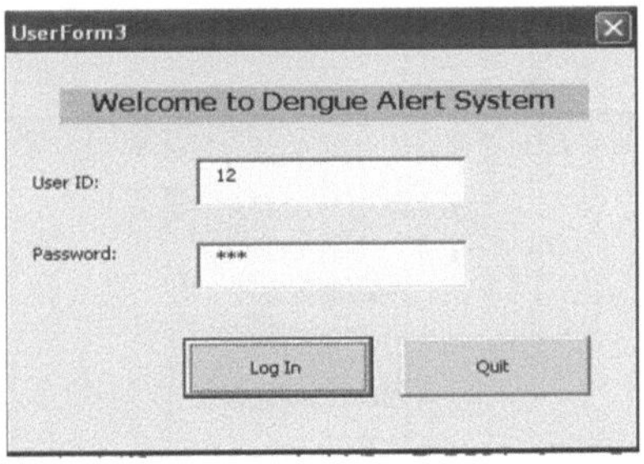

Fig. 8. Login dialog box

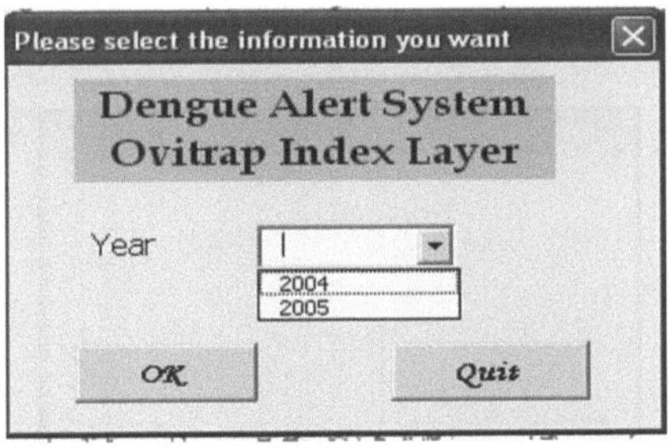

Fig. 9. Selection of temporal data

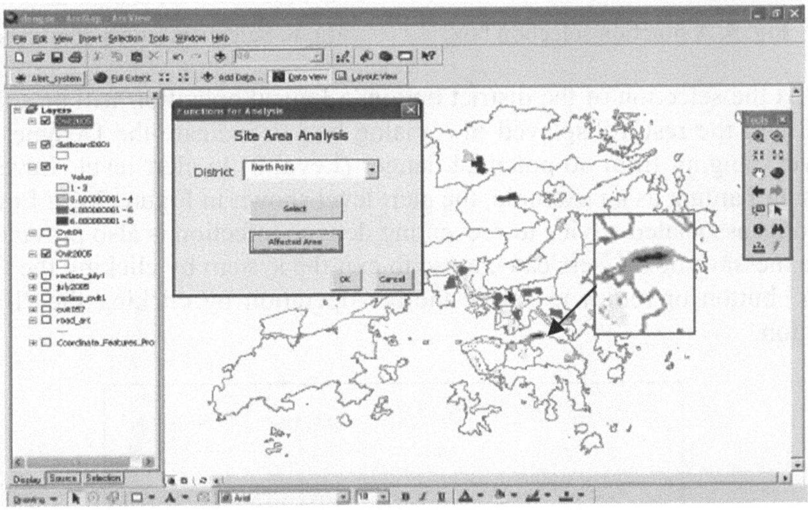

Fig. 10. Site area analysis

Technical users have the option to draw buffer zones around areas of concern. The buffer function delineates clearly affected and potentially affected areas (see enlarged section in Figure 10). The professionals responsible for control measures have the flexibility to expand the buffer distance to explore potential impact over a wider area and estimate the resource implications.

3.3. Analysis Process

3.3.1. Weighted Overlay

The weighted overlay procedure is available for Level 3 users expected to have operational proficiency even without the aid of a user interface. They are given access to the databases of temperature and the Ovitrap Index, along with a base map of Hong Kong.

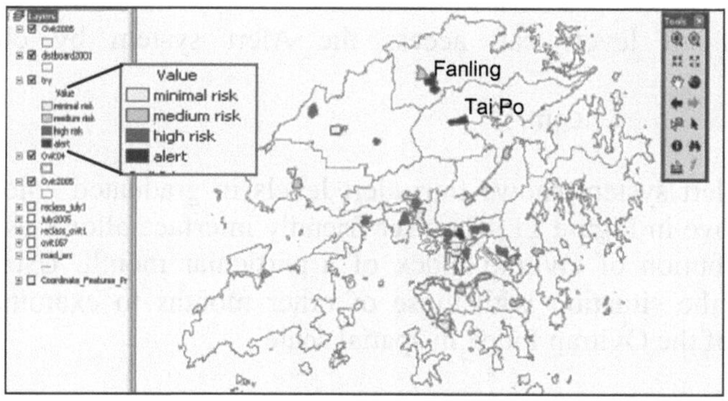

Fig. 11. Results of the weighted overlay

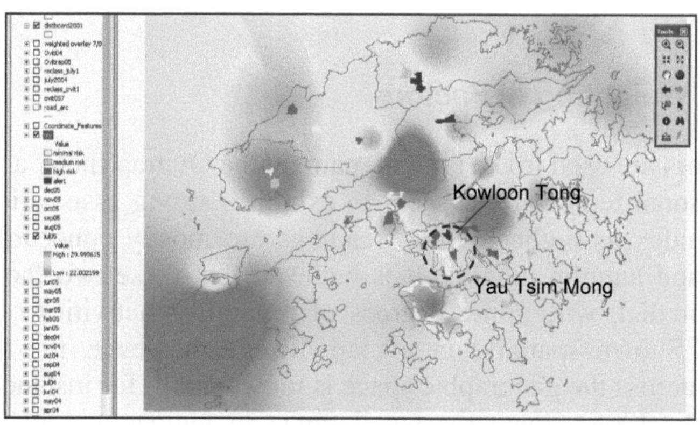

Fig. 12. An overlay with temperature and results of the weighted overlay

Figure 11 illustrates the result of a weighted overlay combining temperature (40% weighting) and the Ovitrap Index (60% weighting) for July 2005. Four alert levels from minimal risks to alert are available and the darker shadings are indicative of areas with higher risks. Thus, Tai Po and

Fanling are classified as the alert areas, as supported by a higher reading in both temperature and the Ovitrap Index in July 2005. When the weighted overlay is superimposed on the temperature layer (Figure 12), it is obvious that the contribution of temperature at 40% is not the deciding factor. The Ovitrap Index at 60% is the overriding factor, resulting in areas with similar temperatures such as Kowloon Tong and Yau Tsim Mong to register a different risk level.

3.3.2. Alert System

Users of all levels can access the Alert system by clicking the

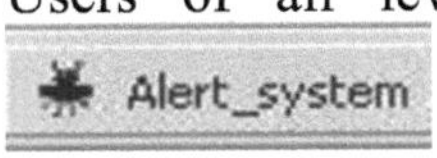

icon.

The Alert system shows four alert levels in graduated colors as illustrated above in Figure 11. The user friendly interface allows users to view the distribution of Ovitrap Index of a particular month. Users can also compare the situation with those of other months to examine temporal changes of the Ovitrap Index in spatial scale.

4. Discussion

4.1. The Weighting of Factors

Two factors are used in the system, namely the Ovitrap Index and temperature. Seasonal temperatures seem to exhibit a strong association with the Ovitrap Index as higher ovitrap readings are mostly concentrated in the summer and autumn months. It is believed that these two factors can be used to predict, with some degrees of reliability, activities of *Aedes albopictus*. Shown spatially in a map display, however, the temperature variation across the geographic space is rather small; for instance, around 2 $^{\circ}$C, in July 2005. Given the small range in temperature by geographic space, it is advised the Ovitrap Index be assigned a heavier weighting to draw out the spatial difference.

4.2. Other Possible Factors

Our system assesses the risk of dengue infection by considering favorable environmental factors for the breeding of *Aedine* mosquitoes and the extent of protection against them. Other potential factors can be added to the system.

Vanwambeke et al. (2006) investigated the determinants of dengue infection in Thailand and discovered that the location of a person and the housing environment were crucial. The density of human population is an important consideration, as a desert region with little or no inhabitants and lots of *Aedine* mosquitoes will not be regarded as risky. Furthermore, housing environments favorable for the breeding of *Aedine* mosquitoes (e.g. hot and humid, with abundant vegetation and open containers, etc.) have a higher risk of dengue exposure. Thus, temperature, humidity, containers, vegetation and topography can be influential factors. Finally, the species of mosquitoes also plays a role, as it is widely known that *Aedes albopitus* may not be the most efficient vector for the virus.

Currently, the FEHD uses the Ovitrap Index alone in the classification of Dengue risk levels. However, the 31 cases of dengue fever reported in 2004 and a similar number in 2005 in Hong Kong showed an even and consistent distribution over the months (Table 2). Evidently, many dengue infections were imported cases, the occurrence of which was unaffected by local risk levels.

Table 2. Dengue Fever Infection (Department of Health 2006)

Disease	Jan	Feb	Mar	Apr	May	Jun	Jul	Aug	Sep	Oct	Nov	Dec	Total
Dengue Fever 2005		2	1	3		2	2	4	5	4	4	4	31
Dengue Fever 2004	5	3	1	1	2	2	3	5	4	1	2	2	31

4.3. Preventive Measures

The main aim of the alert system is to raise public awareness for enhancing personal protection against dengue exposure. Knowledge on the risk of exposure and an advice on appropriate actions are necessarily useful. For members of the public, this alert system must be user friendly, attractive and complete with geographic visualization. The availability of such kinds of information not only allows an individual to take proper actions against dengue exposure in an informed manner but also deters needless speculations.

5. Limitations and Future Prospect of the System

Our system is not without its limitations. In real life, many factors affect the risk of dengue infection, including those of the host, the virus, vector, and the environment. The combination of temperature and the Ovitrap Index offers one aspect of risk assessment, focusing on the determination of the exposure risk alone. Additional factors could be explored and included to enhance the robustness of the system. Besides, other relevant information such as the epidemiology of infections in a region, vector characteristics, and the distinction between imported and local infections, are needed to better interpret the risk levels as defined by the FEHD. The limitations of such an alert system as a warning mechanism on the risk of local exposure should be made known to the population at large.

This pilot project has the potential of expansion to become a real-time system with collaboration from concerned government departments. In the future, it will be meaningful to integrate the system with on-going activities on the prevention and control of dengue fever and to develop training to professionals on dengue vector control. Similar systems can be adapted to cater for other infections such as bird flu, for which environmental factors are important in the propagation of the disease. Moreover, the alert system can be placed on the Internet for widespread accessibility by the general public.

6. Conclusion

We have designed a prototype system to visualize the spatial distribution of the risk of exposure to *Aedine* mosquitoes in the local setting. The system offers information on the incidence of infection and provides a more objective way of defining risk levels. Overall, the system helps to exemplify the spatial concept of risk levels with the aid of a GIS, much unlike the conventional practice of posting statistics about the Ovitrap Index in tabular formats. This system allows visualization of the Ovitrap Index in association with four risk levels for easy interpretation by non-technical users. For researchers with more sophisticated technical skills, dengue fever can be analyzed in both spatial and temporal manner through the weighted overlay analysis in a GIS, thereby enabling an integration of public health matters and geographical approaches.

This project introduces a new method for expressing alert warnings in the spatial scale. The use of spatial graphics provides a clearer picture and an easy interface to better understand the severity or spread of dengue infections, which is considered more superior to data presented in statistical and textual formats. In addition, the weighted overlay method allows the users to explore contribution of factors in varying degrees in determining the exposure risk to dengue. The system therefore serves as both a more objective platform for informing risks and an experimental tool for evaluating the influence of various factors on the risk levels.

It is anticipated that the alert system can contribute to the prevention of dengue infections in real situations. The system can be an effective means of raising awareness of the public in dengue risks because of its direct links to localized advice on preventive measures. However, an effective prevention will require active participation by all sectors of the community including companies and individual families aiming at reducing the breeding grounds of *Aedine* mosquitoes at the source. Disease prevention is a complex problem requiring full community participation.

Acknowledgement

We thank the Food and Environmental Hygiene Department for the provision of Ovitrap Index data for 2004 and 2005. The monthly mean temperature data for 2004 and 2005 came from the website of the Hong Kong Observatory.

Reference

[1] Department of Health, HKSAR. *Monthly Summary Tables of Notifiable Infectious Diseases.* http://www.info.gov.hk/dh/publicat/index.htm Accessed on May 17, 2006

[2] Food and Environmental Hygiene Department (FEHD), HKSAR *Safe Food & Public Health.* http://www.fehd.gov.hk/safefood/dengue_fever/index.html Accessed on May 17, 2006

[3] Vanwambeke, SO et al. (2006) Multi-level analyses of spatial and temporal determinants for dengue infection. *International Journal of Health Geographics* 5:5

GIS Initiatives in Improving the Dengue Vector Control

Mandy Y.F. Tang and Cheong-wai Tsoi

Hong Kong SAR, China

Abstract: The Hong Kong Government has been conducting a dengue fever surveillance program using ovitraps for several years. In the current practice, the disease patterns and vector trends are mapped with the use of a Geographical Information System (GIS) and analyses are then performed to identity buildings having potential danger of a higher Aedes Albopictus infestation. This chapter has two objectives: (1) to review the methodology adopted in this control in Hong Kong, and (2) to propose a GIS based approach to enhance the management and monitoring of the disease, with consideration of temporal, environmental and climatic factors. This study could be important to local communities and countries using ovitraps as a means of controlling dengue infection.

Keywords: GIS, health, Dengue fever, ovitraps, temperature, rainfall

1 Introduction

Dengue viruses have been recognized in recent decades as one of the most dreadful diseases worldwide. Dengue fever is a life threatening disease if a person is being infected more than once. Hong Kong recorded 24 imported cases and 20 local cases of dengue fever in 2002 and a total of 49 cases in 2003 [1]. Although the number of cases is not high compared with some tropical countries, preventive measures and surveillance strategies against the sporadic occurrences of dengue fever are still necessary, especially in a densely populated international city like Hong Kong.

Hong Kong is a Special Administrative Region of China. It borders the South China Sea and Southern China with geographic coordinates at 22° 15 N, 114° 10 E. According to the Hong Kong Observatory, the climate of Hong Kong is subtropical with distinct seasonal variations. In spring (March to mid-May), the temperatures range from 18 °C to 27 °C and the average humidity is about 82 percent. In summer (late May to mid-September), the temperatures rise to 33 °C and the average relative humidity is 86 percent. In 2004 and 2005, the highest number of Aedes albopictus infestation occurred between May and July which suggested that the density of occurrence of the mosquito population is seasonal.

2 Ovitrap Survey

In an effort to prevent and control dengue fever, the Hong Kong Government has been conducting a dengue vector surveillance program using the oviposition trap (ovitrap) since 2000. The use of ovitraps has been considered an effective way for detecting and investigating the oviposition populations. Currently, there are about 2000 ovitraps installed in 38 different locations to assess the presence of Aedes albopictus infestation. Ovitraps installed to detect the presence of Aedine mosquitoes are mainly located outside built structures like housing estates, hospitals and schools, with less than 5 percent of these ovitraps located inside premises. The ovitraps are checked on a regular basis and the Area Ovitrap Index (AOI) of each area is used to indicate the oviposition response by Aedes albopictus mosquitoes. Control measures are undertaken with reference to readings of the ovitrap index.

To analyze the spatial and temporal patterns of oviposition behaviors, a web-based Geographical Information System (GIS) for dengue fever surveillance equipped with query and buffer analysis functions was developed by the Lands Department in 2004 and 2005. Figure 1 shows ovitraps modeled as points and monitoring areas represented as polygons in the web-based GIS. Spatial analyses can be performed on the ovitrap point layer with associated ovitrap count attributes. The system enables officers to query and identify buildings within 100 meter radius of ovitraps indicating the presence of either Aedes albopictus larva or egg (positive count of AOI). These buildings are considered more susceptible to Aedes albopictus.

Apart from the spatial analysis function, GIS can further be extended to provide better insights into the local situation of dengue fever. Two additional types of spatial analyses are proposed in this chapter – one at the local or district level while another at the territory-wide level. The identification of hot spots and the most active ovitraps in terms of mosquito activity in selected sites will be discussed in the local analysis. The results of this analysis can help identity areas of high or low Aedes albopictus population density in the vicinity of particular monitoring areas. The territory-wide analysis makes use of rainfall and temperature data to explore the relationships between climatic factors and the occurrence of Aedes mosquitoes in Hong Kong.

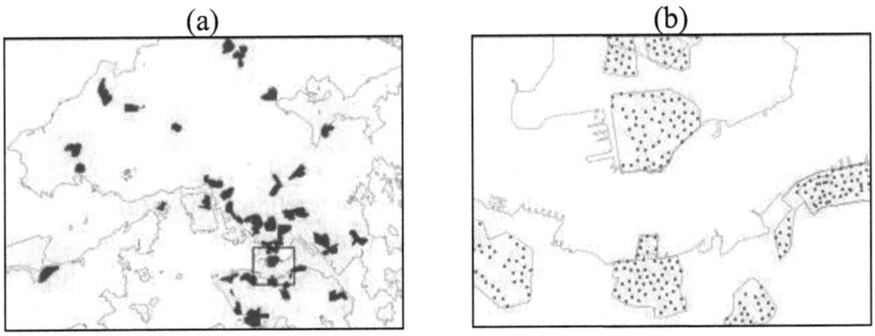

Fig. 1. A distribution of (a) 38 selected ovitrap sites in Hong Kong and (b) Ovitraps in some sites

3 GIS Analysis

The development of a GIS has enabled epidemiologists to include more easily the spatial component in epidemiologic studies [7]. The reason to employ a GIS is its ability to map health-related information in the geographical context. This spatial and visual platform allows relevant authorities to make better decisions about areas needing more resources and to formulate prevention strategies. The potential of using a GIS in health applications can be illustrated in the paragraphs below.

3.1 Local Analysis

The local analysis focuses on a spatial analysis of the Aedes albopictus mosquito count of individual ovitraps in a particular area and their relationships to the nearby ovitraps. According to the Department of Food and Environmental Hygiene of Hong Kong, Aedes albopictus mosquitoes are weak fliers with an average flight distance of about 100m [3]. For practical purposes, ovitrap data collected in monitoring regions in Hong Kong of over 100m apart should therefore have very little or no direct relationship or implications to each other. Hence, the 38 selected regions for vector surveillance are considered as 38 separate areas for local analysis. The results from this local perspective can facilitate investigation, detection, planning and more target specific responses.

Mappings of health-related data have dedicated mostly to detecting hot spots, i.e. areas of higher risks or concentration of disease occurrences. The definition of a hot spot varies from application to application. In general, a hot spot refers to an area where there is sufficient coming together of certain activities (e.g. Aedes albopictus infestation incidents in this case) such that it is labeled as an area of heavy concentration [2]. With reference to ovitrap data in June 2005, a hot spot analysis was carried out and the results in two areas are as shown in Figure 2. It can be seen that there are several hot spots distributed within the area. The local trends of hot and cold spots can be visualized and identified easily with a continuous surface (Figure 2b). The irregular distribution of the mosquito population throughout a local region obviously is suggesting environmental differences within the area. Small containers, flower pots and discarded tires which are known to be highly suitable breeding grounds for Aedes albopictus might be a cause factor. The hot spot analysis can be performed monthly in a particular region to understand the spatial distribution of hot and cold spots.

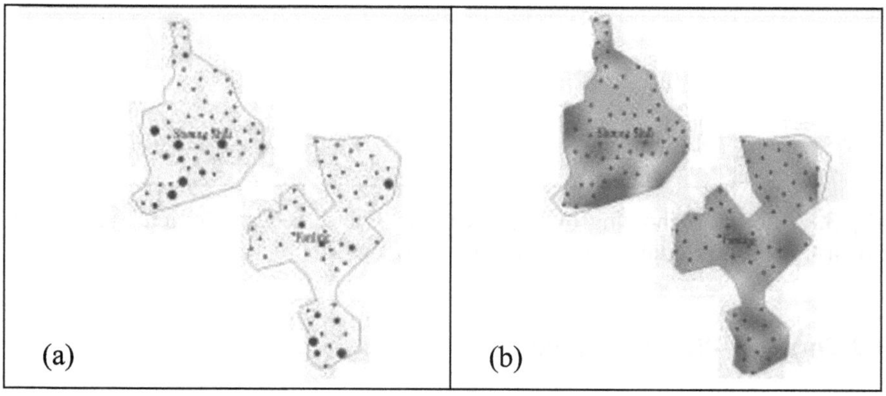

2. (a) Hot spots in Fanling and Sheung Shui, and (b) a continuous surface of a generalized view of hot and cold spots.

In a GIS, spatial statistical analysis can be applied to determine and visualize the magnitudes of reoccurrences of certain events in a particular location. This is useful in studying the conditions of Aedes albopictus mosquitoes and planning control actions in each region. All positive count ovitrap data in 2005 were used. They were converted into weighted point data to extract the most 'active' ovitraps having positive indices over a 12 month period. Figure 3 uses a dot to indicate the location of an ovitrap and the dot size is proportional to the number of incidents at the location. The result reflects that only a few of the ovitraps had positive counts for 6 months or more in 2005. The majority (over 70%) of the devices detected breeding of Aedes albopictus only once or twice a year. This could be the result of community efforts and coordination by the government in environmental cleaning. Based on the results of hot spot and spatial statistical analyses, a list of locations requiring extra efforts to reduce mosquito infestation could be derived. Furthermore, an examination of the spatial distribution of 'active' and 'inactive' ovitraps can pinpoint the most and least likely breeding locations of mosquitoes such that more strategic positioning of ovitraps can be realized.

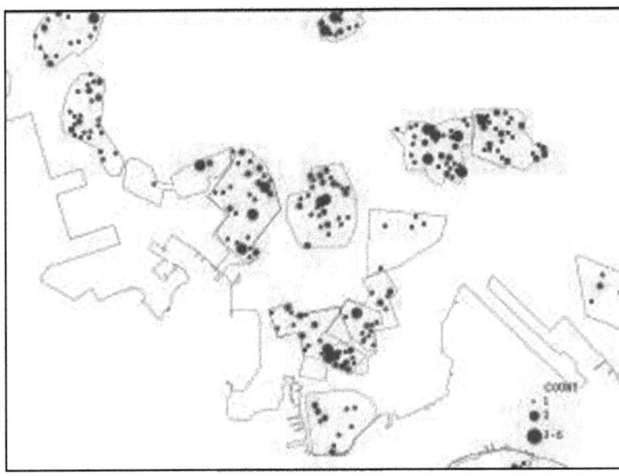

Fig. 3. A dot map showing ovitrap locations where Aedes albopictus are found. Dot size is proportional to the occurrence of Aedes albopictus.

3.2 Territory-Wide Analysis

A study of mosquito population in relation to temperature and rainfall is fundamental to mosquito surveillance [5]. Besides performing analyses of individual events at the local level, a GIS can also be used to investigate the correlation between ovitrap data and various meteorological factors such as rainfall and temperature. Territory-wide analyses of ovitrap data against temperature and rainfall data of 2004 obtained from the Hong Kong Observatory Department were conducted using a GIS. The correlation between rainfall and the positive breeding of Aedes albopictus was computed using the conventional correlation and regression techniques. A regression line was drawn and the Pearson's correlation coefficient (r) was calculated.

In 2004, the regression line representing ovitrap count (X) against rainfall (Y) was $Y = 9.404X + 43.25$ with a correlation coefficient $r = 0.6559$

at the $p <= 0.022$ significance level, reflecting that the breeding of mosquitoes was significantly affected by rainfall in Hong Kong. The regression line of ovitrap count (X) against temperature (Y) was represented by $Y = 0.3595X + 19.53$ with a correlation coefficient $r = 0.8279$ at the $p <= 0.0016$ significance level, revealing a strong association between temperatures and the breeding of mosquitoes. Similar situations were observed using the 2005 data.

The GIS technology offers a completely different way of analyzing and visualizing relationships between spatial and non-spatial features as well as other external factors. The peak period for mosquito breeding was recorded between May and September in 2004. The analysis below uses dengue and meteorological data collected in May, July and September to highlight association between the two variables over different time periods. To start with, all the positive ovitrap data were extracted for cluster analysis. Clusters are deemed to occur in a geographic distribution when groups of discrete features with similar attribute values are found together [4]. In this study, the geographic locations of features (i.e. ovitraps in this study) were used to determine the distribution of clusters. The relationship between mosquito activity and climatic factors can be determined by overlaying the clusters with surfaces representing the temperature gradient and the rainfall distribution (Figure 4).

Figures 4(a) to 4(c) shows the positive ovitrap clusters in relation to rainfall in three different months. It can be seen that most of the clusters were located in regions where rainfall was intense. However, it should be noted that there were some exceptional clusters in areas with a relatively low rainfall. Such a phenomenon could be the result of local human activities or local environmental issues. Referring to the statistical results, there should be a strong correlation between temperature and mosquito activities. However, as shown in Figures 4(d) to 4(f), many clusters were formed in regions with lower temperatures (i.e. areas shaded in lighter tones). This contradiction could be explained possibly by the locations of several of the 31 automatic weather stations in Hong Kong. While the majority of these stations are installed at the ground level of urbanized areas, 5 stations are located at the top of mountains in relatively remote areas (e.g. Tai Mo Shan, the highest mountain in Hong Kong). To understand effects of these remote weather stations in high grounds on the analytical results, another attempt to create the temperature gradient surface was made by isolating the three stations from the dataset: Tai Mo Shan (945m above mean sea-level (msl)), Tate's Cairn (575m above msl) and Nei Lak Shan (747m above msl). It can be seen from Figure 5(d)-5(f) that the positive ovitrap clusters basically coincide with regions of higher temperatures upon the exclusion of the three outlier stations.

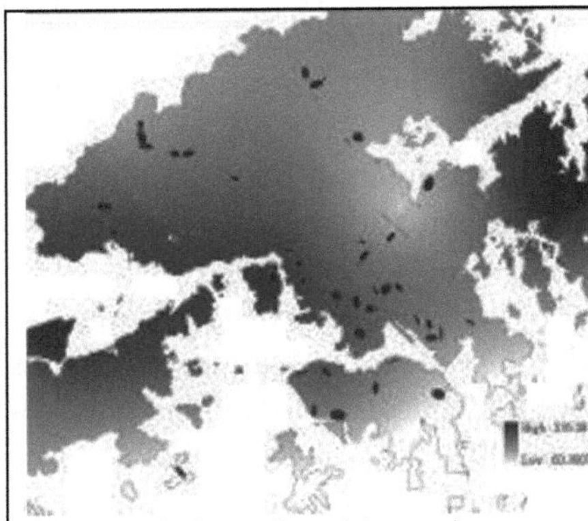

(a) Correlation of positive ovitrap clusters and rainfall in May 2004

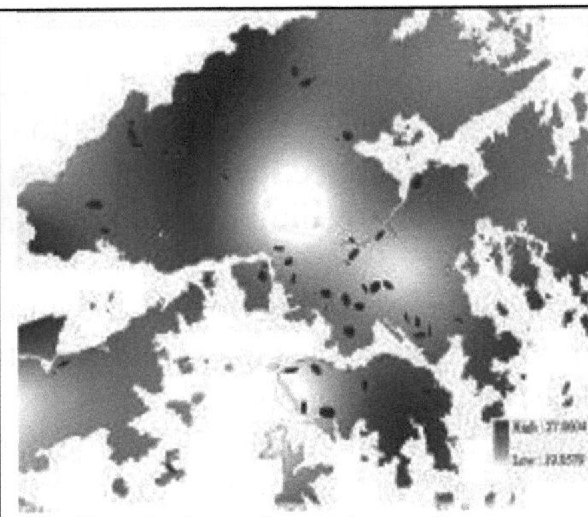

(d) Correlation of positive ovitrap clusters and temperature in May 2004

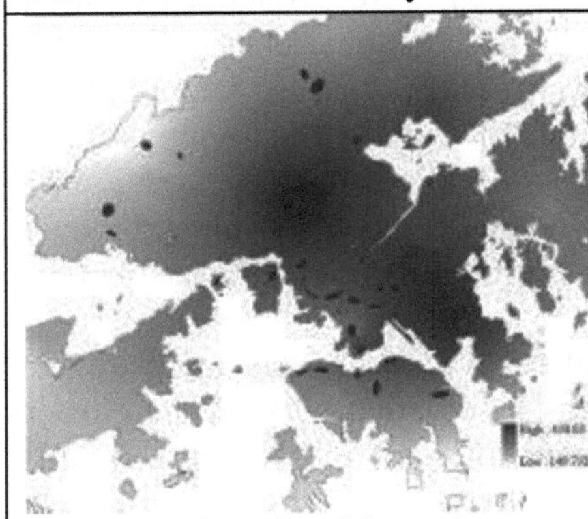

(b) Correlation of positive ovitrap clusters and rainfall in July 2004

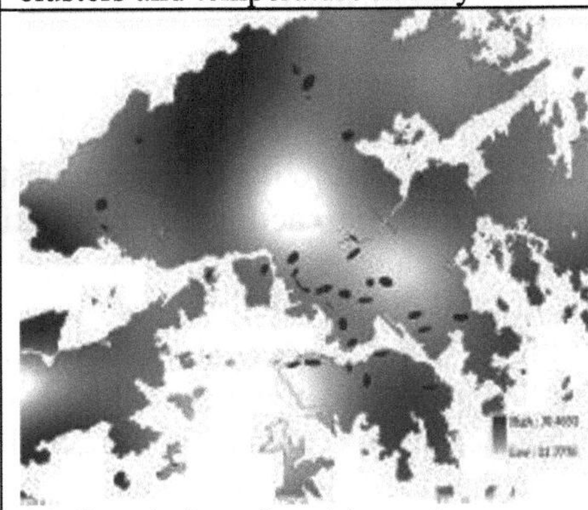

(e) Correlation of positive ovitrap clusters and temperature in July 2004

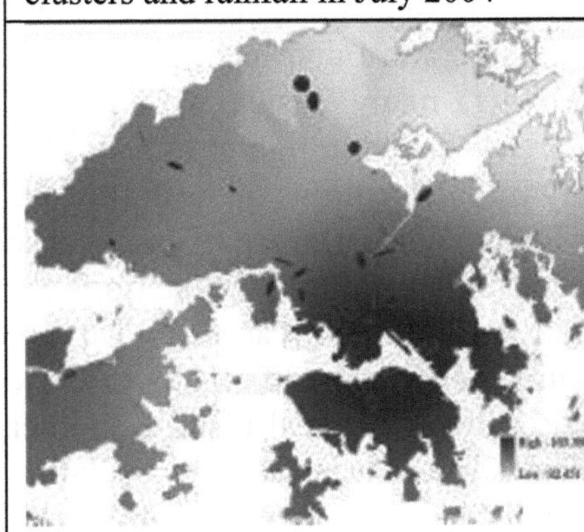

(c) Correlation of positive ovitrap clusters and rainfall in Sept 2004

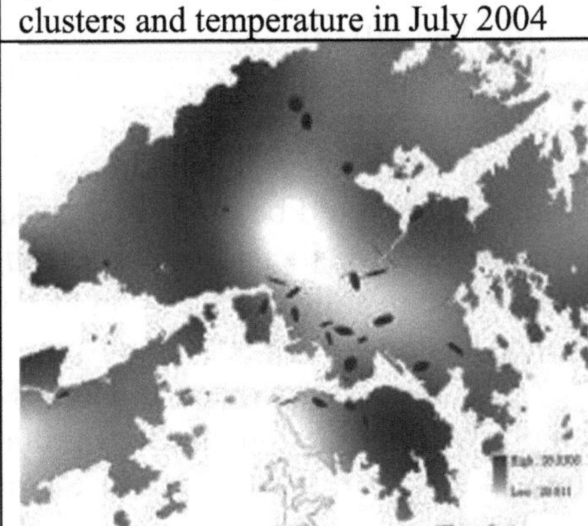

(f) Correlation of positive ovitrap clusters and temperature in Sept 2004

Fig. 4. Maps showing the relationships between areas with a positive ovitrap count against rainfall or temperature

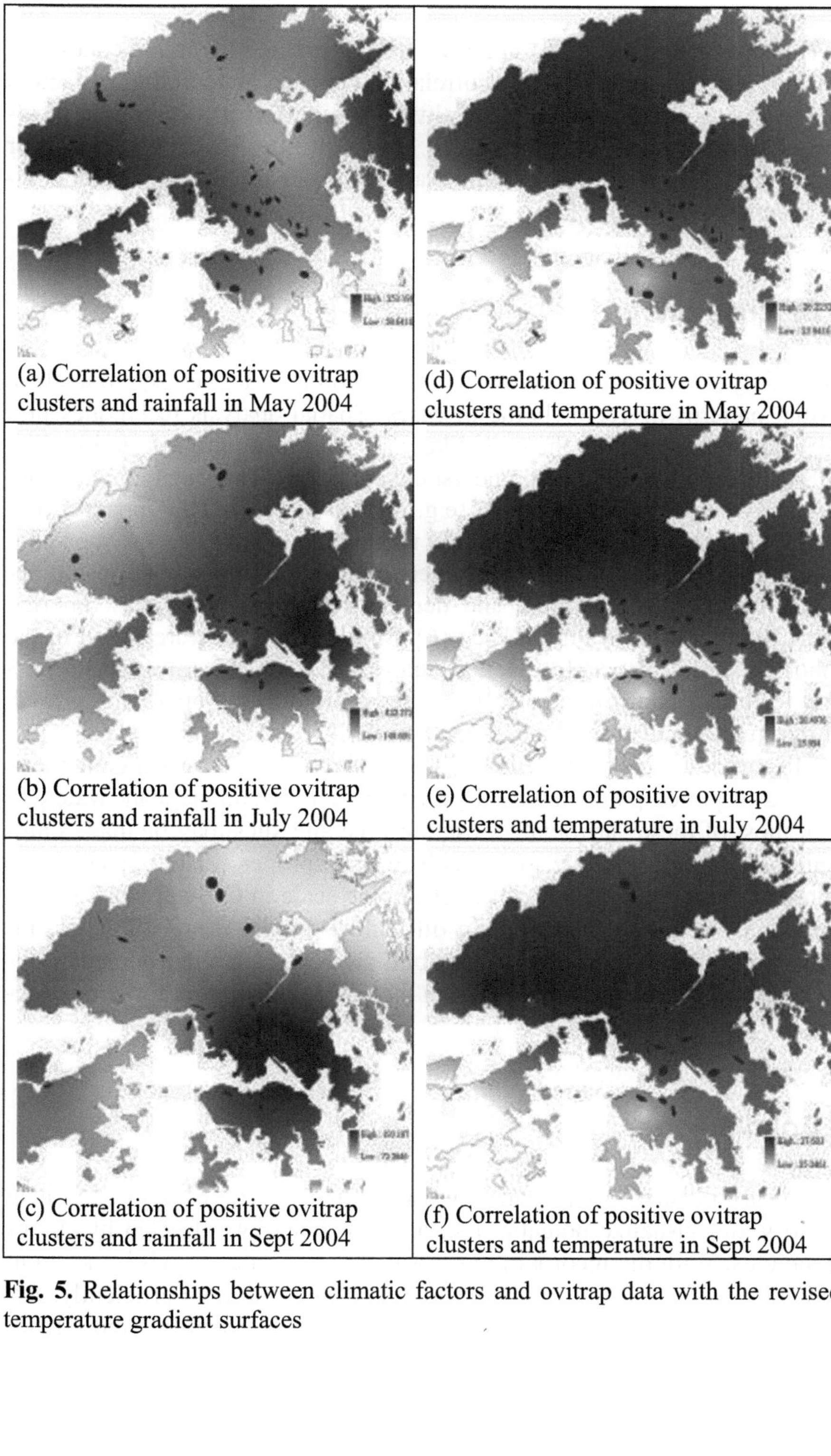

(a) Correlation of positive ovitrap clusters and rainfall in May 2004

(d) Correlation of positive ovitrap clusters and temperature in May 2004

(b) Correlation of positive ovitrap clusters and rainfall in July 2004

(e) Correlation of positive ovitrap clusters and temperature in July 2004

(c) Correlation of positive ovitrap clusters and rainfall in Sept 2004

(f) Correlation of positive ovitrap clusters and temperature in Sept 2004

Fig. 5. Relationships between climatic factors and ovitrap data with the revised temperature gradient surfaces

Higher temperatures and excessive rainfall provide favorable conditions for mosquito breeding. But the correlation between meteorological factors and the mosquito population is not always straightforward and might be affected by other underlying factors [6]. Public utilities like household water supply and storage as well as solid waste disposal are also important environmental determinants [1].

4 Discussions

This study has demonstrated that the GIS approach can be used to analyze spatial patterns of vector borne diseases. Spatial analysis techniques in a GIS can help determine the most likely areas of mosquito infestation. The GIS techniques can also help understand correlations between climatic factors and vector surveillance data (e.g. ovitrap indices). The impact of climatic factors such as temperature, rainfall and humidity on mosquito breeding should not be belittle or oversimplified. A weak relationship between the mean temperature and dengue vector incidence was reported from 1988 to 1992 in Puerto Rico [6]. Similar findings were found in the Johor State of Malaysia [6]. Hence, analysis of a longer term is still necessary to further confirm the effects of meteorological factors on the growth of mosquito population.

Mosquitoes usually breed in pails, watering cans, discarded containers, saucers of potted plants, and gully traps which contain stagnant water. In fact, most of these "artificial" mosquito breeding sites may be made unintentionally by people due to inadequate or improper waste management. Some of these containers are of microscopic scale in GIS applications and their dynamic nature makes them difficult to model in a GIS. Thus, involvement of the community and public awareness on environmental hygiene are of paramount importance in the control of vectors for dengue fever .

5 Conclusions

This study illustrates that spatial statistical analysis can be applied to help identify and visualize potential risk areas or hot spots affected by Aedes albopictus. With the incorporation of surveillance data in a GIS, mosquito activities can be tracked and monitored regularly and more efficiently. Areas repeatedly having positive ovitrap counts can be identified more readily to target more focused cleaning activities. Furthermore, the study illustrates that rainfall and temperatures are two factors among others that affect the breeding of mosquitoes. Without a doubt, a GIS is seen as a useful tool that assist in decision making, adjusting monitoring scheme, deploying resources and coordinating cleaning efforts – all of which are essential elements to controlling the spread of dengue fever in local communities. But community involvement and awareness are also important in minimizing breeding grounds and the occurrences of mosquito infestation which, in turn, contribute to the success of the dengue vector surveillance and control programs.

Acknowledgements

The work described in this chapter was substantially supported by the development team of the "HKSAR Geospatial Information Hub (GIH)" in the Lands Department and ovitrap data from Risk Assessment and Communication Division in the Food and Environmental Hygiene Department. The authors thank both parties.

Reference

[1] Centre for Health Protection, Department of Health, HKSAR (2005) A Three-Year Strategic Plan for the Prevention and Control of Dengue Fever in Hong Kong Report of the Scientific Committee on Vector-borne Diseases

[2] CrimeStat Manual, Chapter 6 – 'Hot Spot' Analysis I http://www.icpsr.umich.edu/CRIMESTAT/download.html Accessed on April 5, 2006

[3] Food and Environmental Hygiene Department – Dengue Fever http://www.fehd.gov.hk/safefood/dengue_fever/index.html Accessed on April 5, 2006

[4] Mitchell Andy (2005) The ESRI Guide to GIS Analysis vol 2: Spatial Measurements & Statistics. Redlands, Calif: ESRI Press

[5] Schulz Travis, Bichler Mike, Voller Laura, Hoff Jennifer (2003) North Dakota Mosquito Surveillance 2003 Program Division of Microbiology, Department of Health, North Dakota

[6] Seng Su Bee, Chong Albert K, and Moore Antoni (2005) Geostatistical Modeling, Analysis and Mapping of Epidemiology of Dengue Fever in Johor State, Malaysia. Proceedings of the Seventeenth Annual Colloquium of the Spatial Information Research Centre. 24 November to 25 November 2005, Dunedin, New Zealand. University of Otago, pp 109-123

[7] Tran Annelise (2001) Dengue Spatial and Temporal Patterns, French Guiana, 2001. http://www.findarticles.com/p/articles/mi_m0GVK/is_4_10/ai_n6076445 Accessed on April 3, 2006

Remark

The authors work in the Land Information Centre of the Lands Department in Hong Kong SAR, China. The views expressed in this chapter represent the views of the authors and do not necessarily reflect the views of the Lands Department or the Hong Kong SAR Government.

Socio-Demographic Determinants of Malaria in Highly Infected Rural Areas: Regional Influential Assessment Using GIS

Devi. M. Prashanthi[a], C.R. Ranganathan[b] and S. Balasubramanian[a]

[a] Division of Remote Sensing and GIS, Department of Environmental Sciences, Bharathiar University Post, Coimbatore, Tamil Nadu, India
[b] Department of Mathematics, Tamil Nadu Agricultural University, Coimbatore, Tamil Nadu, India

Abstract: Geographical Information Systems for disease surveillance play a major role in public health and epidemiology. Malaria is a serious tropical disease and micro level mapping of the disease and its determinants are necessary. Several studies have identified that the major malaria vector Anopheles breeds in pools and streams and hence the people living in close proximity to such breeding sites are at a high risk. The impact of topography and human activities on the incidence of malaria was investigated in a longitudinal trial study conducted in the highly endemic Vellar region of Salem in India using a Global Positioning System (GPS) and a Geographical Information System (GIS). A group of highly endemic villages and an uninfected village with similar topography were surveyed for their geographical features, distance from house to breeding sites, the socio economic living condition of a community (including the construction type of houses), and the use of anti-malaria measures. The results indicated significant relationships between the disease risk and the livelihood patterns of the infected community like sanitation, livestock dependence, etc. A comparative study against the uninfected village highlighted heterogeneous differences among the regions. Based on the integration of social, biological and geographical sciences, the present study provides the regional malaria control authorities an opportunity to assess the risk of encountering the disease infection and to plan prevention measures accordingly.

Keywords: Malaria, disease surveillance, logistic regression

1 Introduction

Natural ecosystems throughout the world have been severely altered by human intervention (Tadei et al. 1998). The rapid urbanization in many parts of the world is changing the context for human population and their interaction with the natural ecosystem. To understand the complex nature of the mosquito human relationship, it is required to identify the type of human migration, population growth, socio economic status, behavior and the environmental aspects around them. This requirement underscores the importance of human intervention that may affect the mosquito vector population and the intensity of parasitic transmission in endemic areas, whether in rural or urban settings. The key determinants of the outcome of malaria should be related to the human host, parasite, vector or environmental parameters. However, the relative importance of these factors has yet to be determined (Greenwood et al. 1991; Marsh 1992; Mbogo et al. 1999).

The habitats of the Anopheline vector and its ecological settings in rural areas have been well documented by earlier studies (Minakawa et al. 1999; Gimnig 2001; Chinery 1984; Barbazan 1998). Vector densities are normally higher in rural than urban areas due to favorable habitats (Coene, 1993). Poverty, farming activities, deteriorating infrastructure and over crowding also contribute to the favorable conditions for developing mosquito breeding habitats (Keating et al. 2003). If households in a community are unable to protect themselves, the risk of transmission of the disease increases. Earlier studies have focused on identifying potential factors in rural areas that help the transmission of the disease such as the type of house construction (Konradsen et al.2003; Hoek et al. 2003), socio economic status of the public (Charlwood et al. 2003; Singh 2003), livestock dependence (Mbogo et al. 1999), behavioral aspects among the residents (Roper et al. 2000; Keating et al. 2003), drainage (Keating et al. 2003; Eisele et al. 2003), distance to streams (Mbogo et al. 1999; Konradsen et al. 2003; Hoek et al. 2003; Staedke et al.2003), and infectious-bite avoidance pattern (Mbogo et al. 1999; Roper et al. 2000; Konradsen et al. 2003).

Understanding the influence of human activities on population dynamics of the malaria vector is important in identifying areas in need of the source reduction efforts and further control. To identify these factors, a longitudinal study of malaria infected zones was carried out to record the socio economic and the demographic data in highly endemic zones of the Salem district using a questionnaire survey accompanied by a hand held GPS (Garmin III Plus). As all the data were collected with household identifiers, opportunities for examining the spatial hypothesis exist. A map of the households and their related surveyed regions can be linked to existing data sources through a GIS.

2 Study Area

Our study area is rather large and covers an size of 5494 sq. km. Surveying an area of this size is a laborious task; therefore, we have selected a small region within the study area for the epidemiological survey using a questionnaire and the Garmin III plus GPS (Figure 1). The study area chosen for the survey was the Vellar PHC region based on its high infection rate and the Santhaithanapatty PHC region as the controlled area. Three village panchayats were surveyed which had seven hamlets in them. Thoppiar, a major perennial stream, runs through these villages.

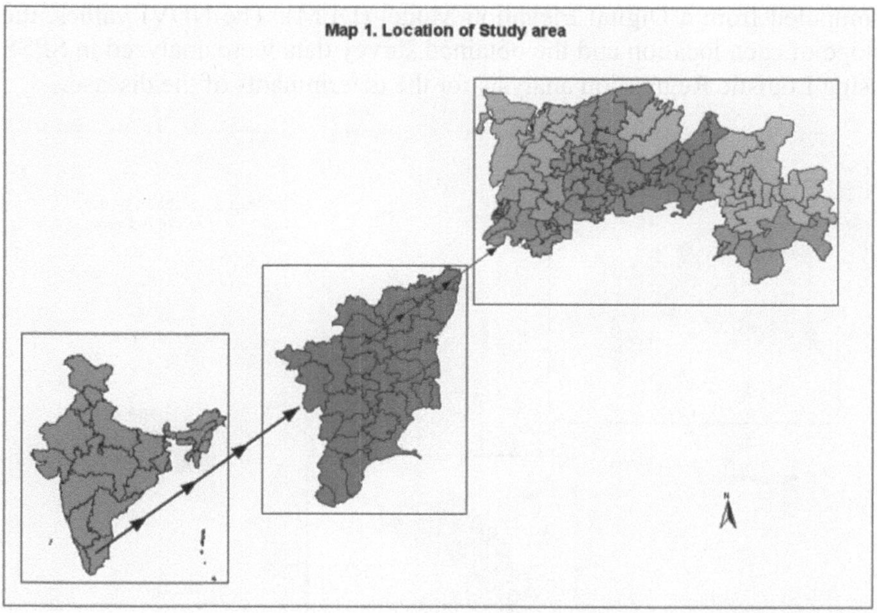

Fig. 1. Location of the study area

Only hamlets that could be visited either by walking or driving tractors were chosen for the study. Each household was visited based on the house number and the questionnaire was filled in by personal interview with the residents. The living habitats of the people, occupational pattern and basic village amenities were also enquired about. The distance from the hamlet to the village hospital was also measured.

3 Methodology

The survey was carried out during the month of October 2004 at four infected villages that come under the Vellar PHC and one non-infected village under the Santhaithanapatty PHC. The survey questionnaire included queries of socio economic status, environmental conditions, human behavior and possible locations of the breeding sites. Random sampling was done in each village for a total of 206 samples which were geocoded using the Garmin III plus GPS. The surveyed information was recorded into a database and the input of data was facilitated with EpiInfo. The data were later imported into the Arcview GIS 3.2a for further processing. The locations of the selected surveyed areas are presented in Figure 1. The Normalized Differential Vegetation Index (NDVI) of these villages was calculated using the ERDAS Imagine 8.5 image processing software for the subsequent month of Nov 2004 (Figure 2). The slope of the study area was also computed from a Digital Elevation Model (DEM). The NDVI values, the slope of each location and the obtained survey data were analyzed in SPSS using Logistic Regression analysis for the determinants of the disease.

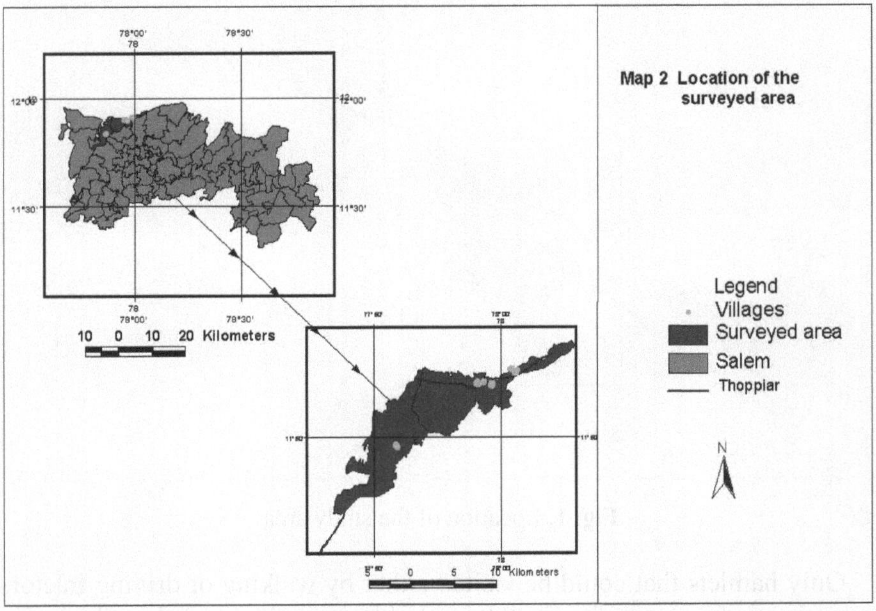

Fig. 2. Location of the surveyed area

4 Statistical Analysis

Logistic Regression is an extension of the multiple linear regression method and it is used when the dependant variable is dichotomous (non–metric) and the independent variables normally are metric. Logistic Regression is also known as Logistic analysis. This technique can be used to describe the relationship of several predictor variables (X1, X2, ... Xk) to a categorical dependant variable Y, typically coded as 1 or 0.

The expected value of Y (E(y)) is in the form of

$$E(y)=\frac{1}{1+\exp-\left[\beta_0+\sum_{j=1}^{k}\beta_j\ X_j\right]} \qquad \text{Eqn (1)}$$

The probability of the expected Logistic value i.e., the probability of occurrence of any of the possible outcome y is rewritten as:

$$\Pr=\frac{1}{1+\exp-\left[\beta_0+\sum_{j=1}^{k}\beta_j\ X_j\right]} \qquad \text{Eqn (2)}$$

The right head side of the equation 2 is the general expression of

$$f(z)=\frac{1}{1+e^{-s}} \qquad \text{Eqn (3)}$$

where,

$$z=\beta_o+\sum_{j=1}^{k}\beta_j X_j \qquad \text{Eqn (4)}$$

The function f(z) is called the logistic function and its values range from 0 to 1 as z varies from -α to +α. This function is widely applied in epidemiological modeling of the probability of a disease.

The general form of the logistic model is

$$\text{Logit } y=\beta_0+\beta_1(X_1)+\beta_2(X_2)+\beta_3(X_3)+\beta_4(X_4)+\ldots+\beta_K(Y_K) \qquad \text{Eqn (5)}$$

A conditional Multiple Logistic estimation was applied in the present study for estimating the parameters (ß) to employ a conditional likelihood function which led to the biased estimates of ß.

Based on the above, a logistic model was proposed to postulate the presence of mosquitoes (absence (0) / presence (1)). The equation (5) is now rewritten for the present model as

$$\text{Logit } y = \beta_0+\beta_1(X_1)+\beta_2(X_2)+\beta_3(X_3)+\beta_4(X_4)+\beta_5(X_5)+ \beta_6(X_6)+\beta_7(X_7)+\beta_8(X_8)+\beta_9(X_9)+\beta_{10}(X_{10})+\beta_{11}(X_{11}) \qquad \text{Eqn (6)}$$

where,

$Y =$ Presence of mosquitoes (No=0, Yes=1)
$X_1 =$ Vegetation (No=0, Yes=1)
$X_2 =$ Slope (No=0, Yes=1)
$X_3 =$ Drainage (No=0, Yes=1)
$X_4 =$ Wood storage (No=0, Yes=1)
$X_5 =$ Sanitation (No=0, Yes=1)
$X_6 =$ Livestock Dependence (No=0, Yes=1)
$X_7 =$ Waterbody (No=0, Yes=1)
$X_8 =$ Education (No=0, Yes=1)
$X_9 =$ House type (Reinforced concrete =0, Tiled=1, thatched =2, Combined=3)
$X_{10} =$ Source of water (Tanks=0 Wells=1, Pipes=2, Open=3, Closed=4)
$X_{11} =$ Prevention measure (No=0, Yes=1)

In multiple linear regression or any other regression models, the regression coefficients βj play an important role in providing information about the relationships of the predictors (independent variables) in the model to the dependent variable. In a logistic regression, the quantification of these relationships involves a parameter called “odds ratio’.

The odds ratio is expressed as

$$Odd = \frac{\Pr(event)}{\Pr(noevent)} = e\beta_0 + \beta_1 X_1 + \beta_K X_K \qquad \text{Eqn (7)}$$

The estimated odds ratio is the ratio of the probability of the occurrence of some events divided by the probability that the same event will not occur. We used SPSS to compute the odds ratio of malaria incidences in the surveyed areas using the aforementioned independent variables.

5 Results

Using the GPS values for the elevation of the surveyed households and contour maps digitized through Arcview GIS 3.2a, a Digital Elevation Model was generated to describe the slopes of the area (Figure 3). The elevation of the area varied from 250 m to 480 m above Mean Sea Level. The slope values were derived from this map.

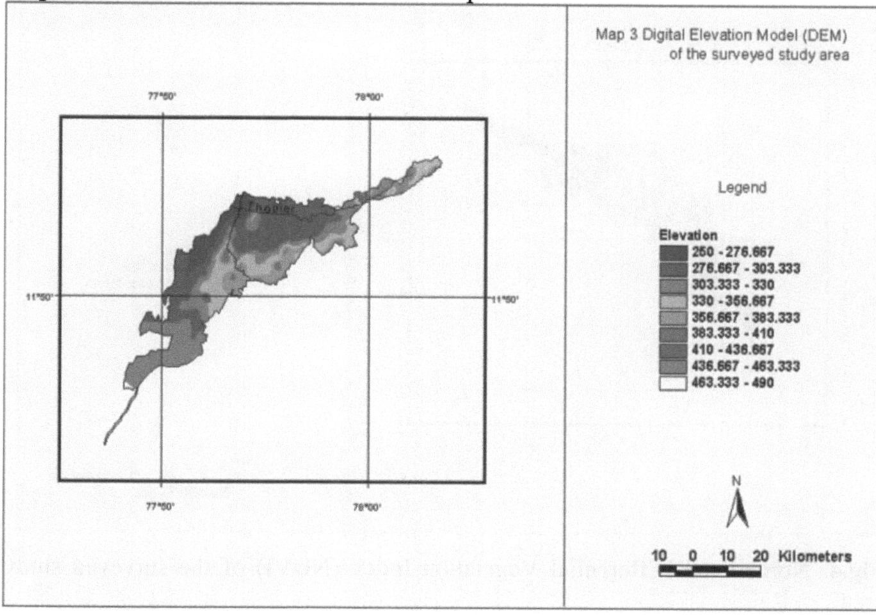

Fig. 3. Digital Elevation Model (DEM) of the surveyed study area

Slope is expressed as the change in elevation over a certain distance. In this case, the certain distance is the size of the pixel. Slope is most often expressed as a percentage but can also be calculated in degrees. The relationship between the percentage and degree expressions of a slope is as follows:

- A 45° angle is considered a 100% slope
- A 90° angle is considered a 200% slope
- Slopes less than 45° fall within the 1 - 100% range
- Slopes between 45° and 90° are expressed as 100 - 200% slopes

For the present study slope was calculated in percentages and vary from 0% to 90%. This slope was attributed to the respective surveyed data to determine the influence of slopes on mosquito breeding sites.

The NDVI, based on spectral reflectance properties, was calculated using the spectral analysis in ERDAS Imagine 8.5 (Figure 4). The NDVI is used to identify the density of vegetation pattern of the selected areas and it was used to recode the data into 3 vegetation classes: densely vegetated area, scrublands or partial fields, and dry lands. These values were attributed to corresponding records in the database.

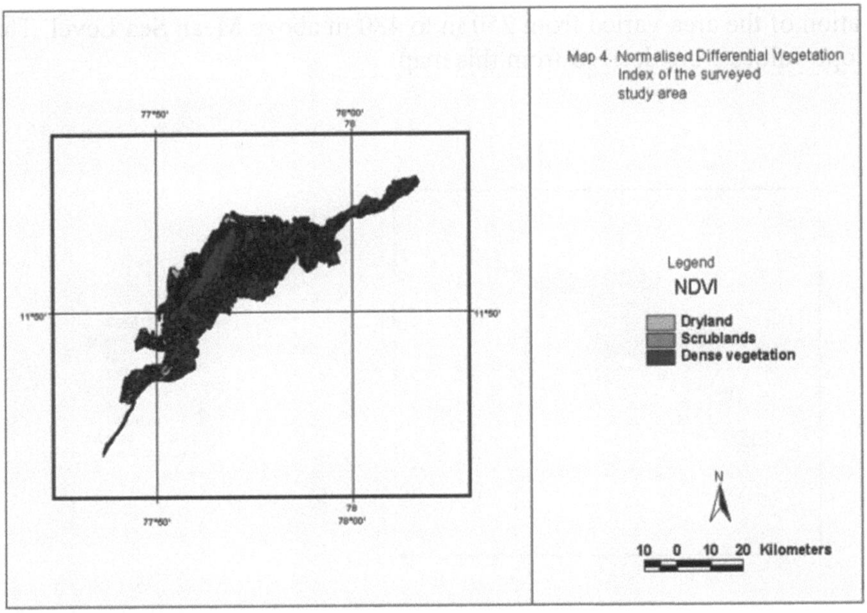

Fig.4. Normalized Differential Vegetation Index (NDVI) of the surveyed study area

A logistic approach was used to identity factors facilitating the proliferation of mosquitoes in the surveyed areas based on the socio demographic data obtained for the infected and controlled surveyed areas. The results are presented in Table 1.

Table 1. The R square values for Model I and II for infected and uninfected areas

Pseudo R-Square	Infected Areas (Model I)	Controlled area (Model II)
Cox and Snell	0.188	0.471
Nagelkerke	0.399	1.000
McFadden	0.326	1.000

Proximity to water bodies or thatched house type were two factors found significant at the 5% level while reinforced concrete houses, tiled houses and the avoidance behavior were significant at the 1% level. In Model II for the controlled area, all the variables were found to be insignificant for malaria proliferation (Table 2).

The fitted model II for the controlled area was

$$\text{Logit } y = \beta_0 \qquad \text{Eqn (9)}$$

Table 2 Logistic Regression analysis of the surveyed variables of infected area.

Variables	Infected Areas					Uninfected Area				
	Coefficients	SE	Wald	df	Signifi-cance	Coefficients	SE	Wald	df	Signifi-cance
Intercept	-17.481	8660.180	0.000	1	ns	176.415	86331.074	0.000	1	ns
Vegetation	-0.247	0.727	0.115	1	ns	-38.003	26432.962	0.000	1	ns
Slope	0.314	0.759	0.172	1	ns	22.859	32593.278	0.000	1	ns
Drainage	1.700	1.171	2.108	1	ns	-39.721	19110.329	0.000	1	ns
[Sanitation=0]	1.057	0.656	2.596	1	ns	-14.531	17043.164	0.000	1	ns
[Sanitation=1]	0.000	0.000	0.000	0	ns	0.000	0.000	0.000	0	ns
[Livestock dependence=0]	0.555	0.660	0.707	1	ns	-56.604	21210.740	0.000	1	ns
[Wbdist=0]	-2.610	0.958	7.425	1	**	20.643	8589.411	0.000	1	ns
[Education=0]	1.302	0.767	2.882	1	ns	0.000	0.000	0.000	0	ns
[Education=1]	0.000	0.000	0.000	0	ns	0.000	0.000	0.000	0	ns
[Housetype=0]	-2.863	1.338	4.579	1	*	-21.508	7829.922	0.000	1	ns
[Housetype =1]	-3.073	1.385	4.925	1	*	-40.296	20631.990	0.000	1	ns
[Housetype =2]	-5.401	2.032	7.065	1	**	-45.660	10740.099	0.000	1	ns
[Source of water=0]	15.029	8660.180	0.000	1	ns	-19.453	17777.925	0.000	1	ns
[Source of water =1]	-2.130	0.000	0.000	1	ns	-18.520	20444.004	0.000	1	ns
[Source of water =2]	-1.717	11290.407	0.000	1	ns	-44.375	0.000	0.000	1	ns
[Source of water =3]	16.008	8660.180	0.000	1	ns	-19.856	39012.161	0.000	1	ns
[Source of water =4]	16.256	8660.180	0.000	1	ns	0.000	0.000	0.000	0	ns
[Prevention =0]	3.483	1.461	5.680	1	*	16.919	15809.724	0.000	1	ns

The following variables were removed from the model: Wood storage, WB dist = 1, Housetype = 3 and Prevention = 1.

6 Conclusion

From the results, it was inferred that among the 206 households surveyed, the mosquito proliferation leading to malaria in the study area was found dependent on the major driving factors identified in Model I. As the main occupation of the people was farming, the economic status of the areas was low and even below the poverty level. The people were found living close to their farms and hence more likely infected due to proximity to stagnant water in the fields and also still water caused by improper drainage that provided breeding grounds for mosquitoes. The houses had mostly thatched roofs. Even in cases of reinforced concrete housing, the adjacent cattle barns were made of thatched roofs, which provided a suitable place for the breeding of mosquitoes. The usage of open ground tanks for water storage afforded another major breeding site. The lack of awareness of proper sanitation facilities and avoidance behavior also added to the context.

In addition, it was noted that people stayed outside their houses most of the time i.e., they slept outside their houses in the evenings or at night when mosquito activity was at its peak. These residents were mostly farmers who wore partial or improper clothing to facilitate the daytime heat. Mostly people reared cattles and other livestocks within their premises which also led to constant and favorable conditions for mosquito proliferation. In the infected areas, the schools were situated closer to farms or had water storage tanks nearby. This practice led to the children being infected more than the adults.

Results of the present study identified target variables that potentially favor mosquito breeding sites in the survey areas. The findings should help in mobilizing social behavior and communication programs that may be implemented to reduce the breeding sites based on community involvement. Additionally, resource allocation can be recommended to infected areas identified through population-based-surveys, which can be conducted as a component of malaria control surveillance.

References

[1] Barbazan P, Baldet T, Darriet F, Escaffre H, Dhoda DH, Hougard JM, (1998) Impact of treatments with B.sphaericus on Anopheles populations and the transmission of malaria in Maroua, a large city in a savannah region of Cameron. J.Am. Mos. Contrl Asso. 14:33-39.

[2] Charlwood JD, Pinto J, Ferrara PR, Sousa CA, Ferreira C, Gil V, Virgillo ER (2003) Raised houses reduce mosquito bites, Malaria Journal 2:45

[3] Chinery WA (1984) Effects of ecological changes on the malaria vectors Anopheles funestes and the Anopheles gambiae complex of mosquitoes in Accra, Ghana . J. Trop. Med. Hyg 87:75-81
[4] Coene J (1993) Malaria in Urban and rural Kinshasa: the entomological input. Med Vet Entomol 7: 127-137
[5] Eisele TP, Keating J, Swalm C, Mbogo C, Githeko A, Regens JL, Githure JI, Andrews L and Beier J C (2003) Linking field based ecological data with remotely sensed data using a geographic information system in two malaria endemic urban areas of Kenya. Malaria Journal 2:44
[6] Gimnig JE, Ombok M, Kamau L, Hawley WA (2001) Characteristics of larval Anopheline (Diptera: Culicidae) habitat in western Kenya. J Med Entomol 38: 282-288
[7] Greenwood BM, Marsh K, Snow RW (1991) Why some African children develop severe malaria? Parasitol Today 7: 277-281
[8] Hoek W, Konradson F, Amersinghe P H, Perara D, Piyaratne M K, Amerasinghe FP (2003) Towards a risk map of malaria in Srilanka: the importance of house location relative to vector breeding sites. Int J Epidemiology 32: 280-285
[9] Keating J, Macintyre K, Mbogo C, Githeko A, Regens JL, Swalm C, Ndenga B, Steinberg L, Kibe L, Githure JI and Beier JC (2003) A geographical sampling strategy for studying relationships between human activity and malaria vectors in Urban Africa Am. J Trop. Med. Hyg 68(3): 357-365
[10]Konradsen F, Amerasinghe P, Hoek W, Amerasinghe F, Perara D, Piyaratne M (2003) Strong Association between House characteristics and malaria vectors in Srilanka Am. J Trop. Med. Hyg 68 (2): 177-181
[11]Marsh K (1992) Malaria- a neglected disease? Parasitol 104: S53-S69
[12]Mbogo CNM, Kabiru EW, Gregory EG, Forster D, Snow RW, Khamala CBM, Ouma JH, Githure JI, Marsh, K and Beier JC (1999) Vector related case control study of severe malaria in Kilifi district, Kenya Am. J Trop Med Hygiene, 60(5): 781-785
[13]Minakawa N, Mutero CM, Githuire JI, Beier JC, Yan G (1999) Spatial distribution and habitat characterization of anopheline mosquito larva in Western Kenya. Am. J Trop. Med. Hyg 61: 1010-1016
[14]Roper MH, Torres RSC, Goicochea CG, Anderson EM, Guarda JSA, Calampa, Hightower A W, Magill A J (2000) The Epidemiology of malaria in epidemic area of the Peruvian Amazon Am. J Trop. Med. Hyg 62(2):247-256
[15]Singh N, Mishra AK, Shukla MM, Chand SK, (2003) Forest malaria in Chhindwara, Madhya Pradesh, Central India: A Case study in Tribal Community Am. J Trop. Med. Hyg 68(5): 602-607
[16]Staedke SG, Nottigham EW, Cox J, Kamya MR, Rosenthal PJ and Dorsey G (2003) Short report: Proximity to Mosquito Breeding sites as a risk factor for clinical malarial episodes in an urban Cohort of Ugandan Children. Am. J Trop. Med. Hyg 69(3): 244-246
[17]Tadei WP, Tatcher BD, Santos JMM, Scarpassa VM, Rodrigues IB, Rafeal MS (1998) Ecologic observations on Anopheline vectors of Malaria in the Brazilian Amazon Am. J Trop Med Hyg, 59(2): 325-335

A Study of Dengue Disease Data by GIS Software in Urban Areas of Petaling Jaya Selatan

Mokhtar Azizi Mohd Din, Md. Ghazaly Shaaban, Taib Norlaila and Leman Norariza

Department of Civil Engineering, Faculty of Engineering, University of Malaya, 50603 Kuala Lumpur, Malaysia

Abstract: The MapInfo GIS software was used to map dengue data and its relationship to the sanitary conditions in the study site of Petaling Jaya Selatan, Selangor, Malaysia. The study utilized dengue data provided by the Petaling Jaya Municipal Health Unit and it was integrated with the Petaling Jaya map to pictorially present information about the affected areas. The software gives a rapid identification of the locations of dengue cases through a graphics display. This multi racial residential area reported 35 cases of dengue over a seven-month study commencing between November 1998 and June 1999. The racial breakdown of patients was 57% Chinese, 20% Indian and 23% Malay among which 54% was female and 46% male. Based on the on-site survey, it was observed that the affected areas have very poor sanitary living conditions. This study concluded that poor sanitary conditions were the most probable cause for the spread of such a disease and the GIS technology has been proven efficient in data collection and presentation of disease incidence for charting immediate corrective and preventive actions.

Keywords: health, MapInfo Geographical Information System (GIS), dengue, *Aedes aegypti*, improper drainage

1 Introduction

Dengue is transmitted by the Aedes aegypti mosquito species found predominantly in urban centers and in certain rural areas of Malaysia. It is rarely observed at an elevation of 4,000 feet above sea level. This species breeds in stagnant water and poorly drained areas.

Dengue fever is characterized by a sudden onset of high fever, severe headaches, joint and muscle pain, nausea, vomiting and rash (National Centre for Infectious Diseases 2000). The rash appears 3-4 days after the onset of fever and the illness may last up to 10 days but complete recovery can take 2 to 4 weeks. There is no known vaccine for dengue fever. Eradication is through preventative methods by instilling awareness on improved public health through education and therefore improvement in general hygiene is most crucial.

Previous work by Md. Ghazaly et al. (1998) on a similar research on water quality and health investigated the impact of water quality and quantity on human health. It was observed that a general improvement in both the quality and quantity of water supplied to users was necessary to curtail the spread of the disease. This was especially true for raw river water quality prior to treatment and post treatment at the water treatment plant. However, no direct link between raw water, treated water at the water treatment plants and the disease could be ascertained due to inadequate information. The importance of sanitation (including proper disposal of wastes) was also emphasized in the prevention of the disease from spreading. At that time, no GIS work was carried out due to the preliminary nature of the work.

2. Role of GIS

The Mapinfo GIS software (MapInfo Corporation 1996) has the ability to present dengue data spatially. Spatial data analysis involves describing data related to a process operating in space, the exploration of patterns and relationships in such data, and the search for explanations of such patterns and relationships (Bailey and Gatrell 1995). Coupled with time based data, useful information can be extracted from locational data and layering of analytical outputs for visual presentation. Data about disease incidence, including location and date of occurrence, can be incorporated easily in a GIS for comprehensive analyses.

In this study, we managed to convert dengue data found in various formats into the MapInfo GIS format. Having arranged the data in a systematic way, the relevant authority was able to retrieve and view the required information efficiently. This information on Dengue prevalence is essential for town and urban planning units in coordinating and managing works in environmental health.

3 Methodology

3.1 Selection of the study area

Petaling Jaya Selatan (PJS) was chosen because it is located in the urban area where dengue cases are frequently reported in comparison to the other areas of Petaling Jaya. Given the scope and time limitation of the study and the availability of surveillance data, the investigation was focused on a very limited area as indicated above (Norariza 2000). A digital map of the study area in Petaling Jaya was obtained from the Jabatan Ukur dan Pemetaan Negara Malaysia (JUPEM).

3.2 Sources of Dengue Case Data

Dengue data was obtained from the Department of Health and Petaling Jaya Municipal Council (Jabatan Kesihatan and Majlis Perbandaran Petaling Jaya) or MPPJ. On-site surveys of the study area were also carried out to identify possible factors to correlate with dengue, such as conditions of sanitation, solid waste disposal and wastewater treatment system.

3.3 Data Input to MapInfo

The addresses of all reported dengue cases were input into the MapInfo GIS. The location of each case was marked on a digital base map to see its distribution within the study area. Other related information such as date, age, gender, race, drainage system, solid waste disposal system, and building type were arranged as attributes to each case in a tabular format.

4 Analysis of Dengue cases

The study adopts a 5-step procedure for map digitizing and analysis. Firstly, the digital map of Petaling Jaya in dxf format was imported to the MapInfo GIS for further processing (Figure 1). Then, Dengue data was aligned with the base map and mapped. The Dengue locations were linked with the corresponding attributes for easy retrieval and display in MapInfo (Figure 2). The map could be enlarged or zoomed in for a better view to permit editing and analysis. Finally, the exact locations of the dengue cases could be studied in relation to other factors.

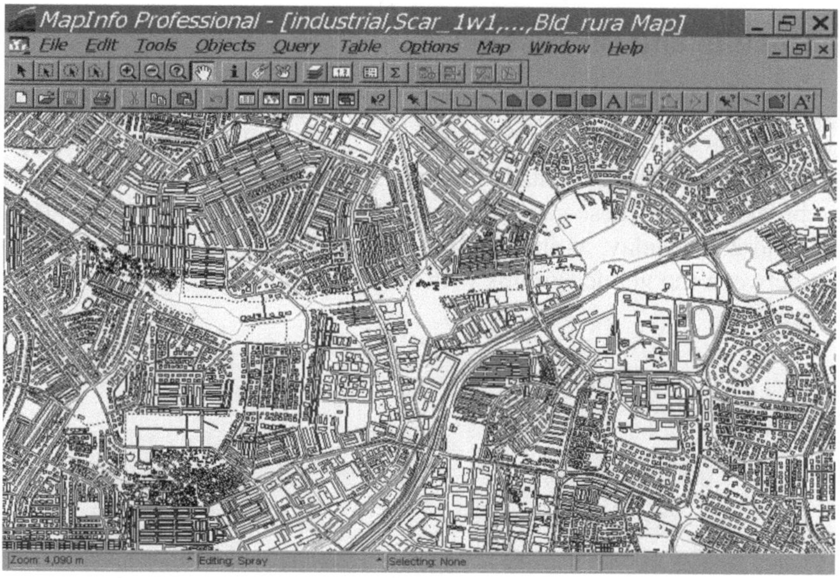

Fig. 1. A digital base map of Petaling Jaya from JUPEM

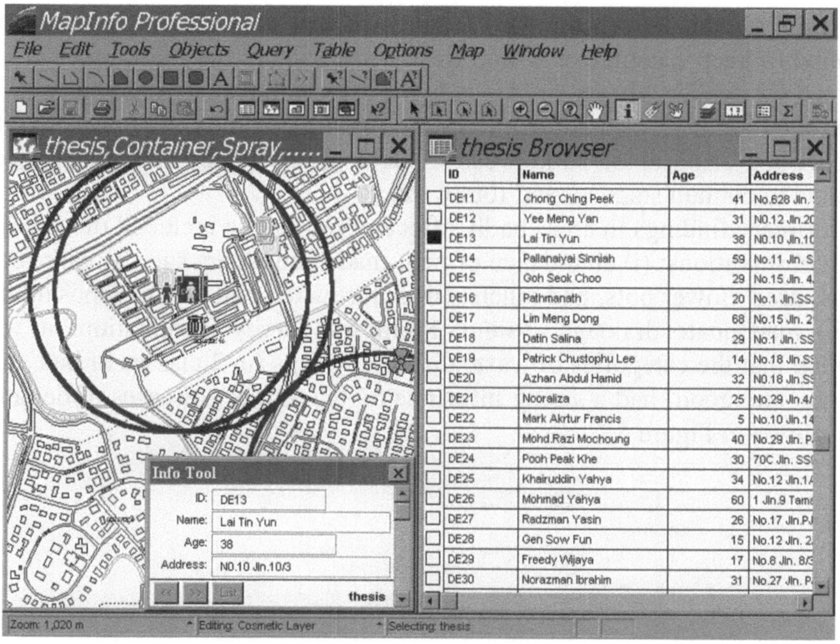

Fig. 2. MapInfo showing attributes of selected cases

5 Results and Discussion

35 cases of dengue were reported over an eight month period from November 1998 to June 1999. The Dengue cases were reported monthly to the hospitals and the patients for the study aged between 4 to 62 years old. The breakdown of these cases by racial groups was 57% Chinese, 20% Indian, and 23% Malay, as indicated in Figure 3. There was an almost equal split between the gender groups with 54% female and 46% male, as shown in Figure 4.

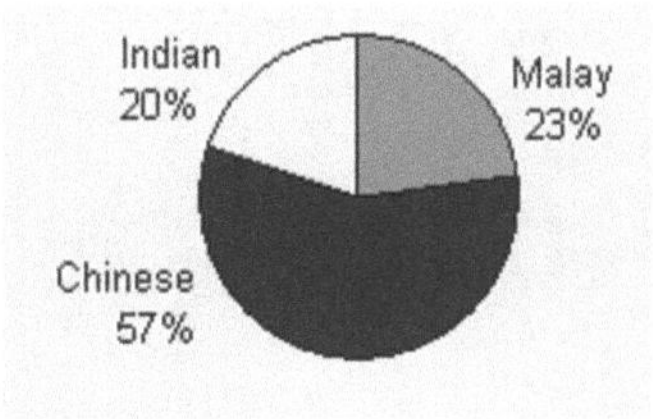

Fig. 3. Dengue case according to race

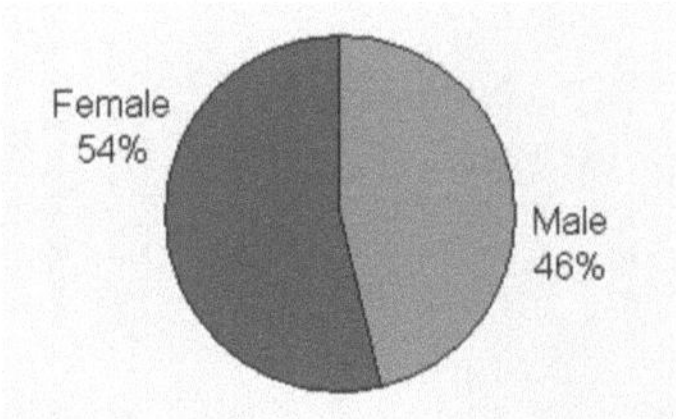

Fig. 4. Dengue case according to gender

Data obtained from the MPPJ indicated that each of the patients was found to keep a container within their house, where the mosquitoes could breed easily. Weekly on-site visits to the affected areas in September 1999 for one month yielded some interesting findings. Some of the areas exhibited poor disposal of solid wastes. Most of the uncovered drains were found to contain stagnant and foul water.

Primary findings in areas badly affected by Dengue included the following observations: (i) abundance of man-made containers such as cans, bottles, tires, flower pots, and others; (ii) inadequate solid waste disposal; and (iii) inadequate drainage system. From the survey information, it was found that the categories of waste disposal ratings (in 3 classes of bad, medium, and good) had a strong influence on the number of cases of dengue, as shown in Figure 5.

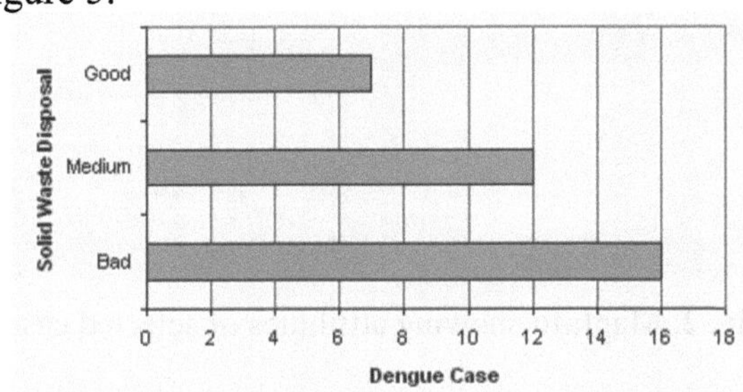

Fig.5. Solid waste disposal versus Dengue case

The Ministry of Health has taken reactive and proactive measures in combating the dengue problems in Petaling Jaya Selatan and particularly in places having a tendency for mosquito breeding. These measures included the following:

- educating the people about the importance of cleanliness and hygiene (Table 1)
- periodic fogging of insecticide such as malathion (Table 2)

Table 1. Public awareness activities by MPPJ

SUBJECT	YEAR				
Health Awareness Education	1993	1994	1995	1996	1997
Health briefing	20	30	95	125	58
Video session	1	1	1	0	0
Demonstrations	0	0	0	0	0
Exhibitions	0	0	0	20	10
No. of posters and flyers	35,000	40,000	70,000	62,000	34,000

Table 2. Fogging activities by MPPJ (Aedes control / Fogging / Health education activities)

Week	No. of Buildings Checked	Building With +Ve Result (Found Aedes)	% OF Building (Found Aedes)	Fogging Activities	Cleaning Activities	Health Briefing
current week	4,066	57	1.4	10,697	0	0
last week	55,853	1,542	2.9	284,479	12	5
total	59,919	1,599	2.9	295,176	12	5

Our study demonstrated that the application of MapInfo in examining dengue data contributed to faster viewing and retrieval of critical health and environmental information, as shown in Figure 6. Appropriate mitigating actions by the relevant authorities can be executed far more quickly once the affected locations are identified.

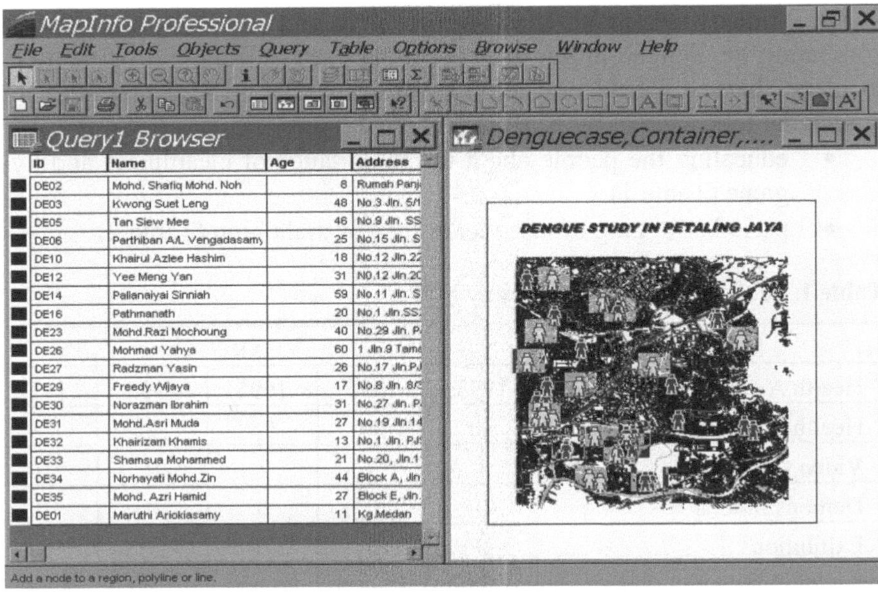

Fig. 6. MapInfo showing a full view of the area studied and patient databases

6 Conclusion

The use of MapInfo GIS in our investigation of the Dengue data yielded some useful information about the disease. On-site visits to the area tended to confirm the data obtained from the Ministry of Health. Factors such as hygienic living conditions, proper disposal of solid wastes and effective drainage play important roles in curtailing the spread of Dengue in the long run. Short term measures of fogging may alleviate the problem to a certain extent. Our future work will involve an exploration of statistical analysis of health related data with environmental and locality factors.

Acknowledgements

The authors would like to thank the Department of Health of Petaling Jaya, Petaling Jaya Municipal Council (MPPJ), Department of Surveying and Mapping Malaysia (JUPEM) and staff of the Department of Civil Engineering, University of Malaya for the technical information supplied and their continuous support.

References

[1] Bailey TC and Gatrell AC (1995) *Interactive spatial data analysis*. 2nd edition, Longman, Harlow
[2] MapInfo Corporation (1996) *Map Info User Manual*, version 4.5. Troy, New York
[3] Md Ghazaly Shaaban, Zubaidah Ismail, Norjidah Anjang Abdul Hamid, Sumiani Yusoff and Nik Meriam Sulaiman (1998) Impact of water quality and quantity on human health. One day seminar on *Recent Advances on Research in the Water Industry*. Malaysian Water Association, 28 July, Kuala Lumpur, pp.1-17
[4] National Centre for Infectious Diseases (2000) *Travellers Health*. Virginia, February
[5] Norariza Leman (2000) Application of GIS in dengue studies in Petaling Jaya. *B.Eng.(Environment) Thesis*, University of Malaya, Kuala Lumpur

A Spatial-Temporal Approach to Differentiate Epidemic Risk Patterns

Tzai-hung Wen[a], Neal H Lin[b], Katherine Chun-min Lin[c], I-chun Fan[a], Ming-daw Su[d] and Chwan-chuen King[b]

[a] Centre for Geographical Information Science, Research Centre for Humanities and Social Sciences, Academia Sinica, Taipei, Taiwan
[b] Institute of Epidemiology, College of Public Health, National Taiwan University, Taipei, Taiwan
[c] Department of Public Health, College of Public Health, National Taiwan University, Taipei, Taiwan
[d] Department of Bioenvironmental Systems Engineering, National Taiwan University, Taipei, Taiwan

Abstract: The purpose of disease mapping is to find spatial clustering and identify risk areas and potential epidemic initiators. Rather than relying on plotting either the case number or incidence rate, this chapter proposes three temporal risk indices: the probability of case occurrence (how often did uneven cases occur), the duration of an epidemic (how long did cases persist), and the intensity of a transmission (were the case of chronological significance). By integrating the three indicators using the local indicator of spatial autocorrelation (LISA) statistic, this chapter intends to develop a novel approach for evaluating spatial-temporal relationships with different risk patterns in the 2002 dengue epidemic, the worst outbreak in the past sixty years. With this approach, not only are hypotheses generated through the mapping processes in furthering investigation, but also procedures provided to identify spatial health risk levels with temporal characteristics.

Keywords: risk identification, spatial autocorrelation, spatial-temporal analysis, epidemic

1 Introduction

In 1854, John Snow curbed the spread of cholera by identifying through mapping the infected water pump located in the Golden Square of London (Snow 1936). Ever since, mapping of infected cases has become an important means to generating hypotheses, guiding intervening measures, and controlling infectious outbreak. Mapping can help epidemiologists to find the origins of an outbreak, but more importantly, to target high-risk areas and places of strategic importance in disease prevention (Ali et al. 2003). A Geographical Information Systems (GIS) is often constructed under these circumstances: a computer-based system is set up to integrate spatial statistics into maps, which can detect disease clustering and determine relationships between disease rates and relevant geographic locations (Lai et al. 2004, Cockings et al. 2004, Dunn et al. 2001). Recent studies have united temporal factors with spatial analysis and found dynamic epidemic changes in certain time course and in specific locations (Tran et al. 2004, Harrington et al. 2005, Morrison et al. 1998, João et al. 2004, Getis et al. 2003). Unlike the traditional approach, this chapter enhances spatial analysis by using not only the incidence rate. We propose a spatial model that integrates temporally-defined epidemiological characteristics and verify it using data from the 2002 dengue epidemic in Taiwan.

Dengue virus, a member of flaviviruses with four known antigenic distinct serotypes, is transmitted in Taiwan mainly by mosquitoes of species *Aedes aegypti* and *Aedes albopictus*. Few studies have focused on the interplay between the spatial clustering of dengue cases and entomological factors (Ali et al. 2003, Tran et al. 2004, Morrison et al. 1998) or environmental conditions (Alpana and Haja 2001) by using a GIS with spatial statistics. However, spatial analysis is the key to identifying the incidence of dengue fever (DF), dengue hemorrhagic fever (DHF) and dengue shock syndrome (DSS). On the contrary, several studies have the treated temporal issues on dengue transmission (Tran et al. 2004, Harrington et al. 2005, Morrison et al. 1998, Derek et al. 2004). Innovative models of temporal-spatial analysis in disease transmission have proffered, and tested by data of the 2002 dengue epidemic in Taiwan, which was the worst outbreak in both magnitude and severity in the past sixty years. Hopefully this novel approach of disease surveillance can be applied to detect other infectious diseases.

2 Materials and Methods

2.1 Data of Dengue Confirmed Cases

Kaohsiung City and its satellite city of Fengshan, confronted in 2002 the biggest dengue epidemic in Taiwan since 1940s. A total of 4,790 cases were confirmed (representing 85% of total cases found in 2002), including 4,574 dengue fever (DF) cases and 286 dengue hemorrhagic fever (DHF) cases according to WHO's definition (Chao et al. 2004). This study focuses on these 4,790 confirmed cases cohered with positive results in one of the three following laboratory tests: (1) molecular diagnosis (Lanciotti et al. 1992), or (2) serological diagnosis (Shu et al. 2001), or (3) virus isolation (Kuno et al. 1985). Figure 1 illustrates the temporal progression of DF versus DHF cases.

This chapter chooses *Li* - the smallest administrative unit in Taiwan - as the spatial mapping unit, and a 7-day week as the temporal indicator. Most *Li* in urban areas occupies 0.26-0.58 square kilometers and comprises 2,100-5,300 residents or 850-1,600 households. 423 *Li* covered roughly 94.4% of total confirmed dengue cases in 2002. These case data were provided by the Taiwan-CDC and without personal identifiers to protect the privacy of the patients. All cases (including DHF cases) collected in this chapter were summarized at the *Li* scale, instead of by exact addresses.

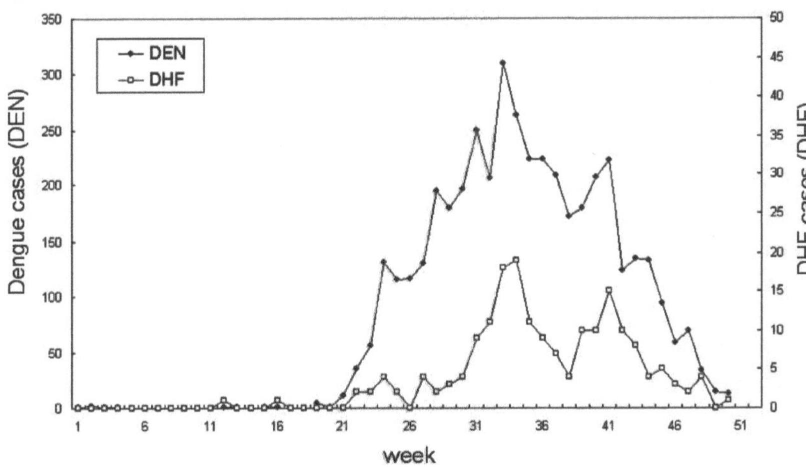

Fig. 1. The epidemic curve of week-based total confirmed dengue (DEN) and dengue hemorrhagic fever (DHF) cases in Kaoshiung City and Fengshan City, 2002

2.2 Temporally-defined Indices as Epidemiological Measures

A risk model developed by Wen et al. (2006) depicts the dynamic process of a dengue epidemic by emphasizing temporal characteristics on map displays. To evaluate a dengue epidemic in both magnitude and severity, this chapter modifies and further introduces indices with temporally-defined epidemiological characteristics (also referred to as *temporal indices*), including (1) probability of occurrence α or how often do cases occur in certain week(s) of a calendar year (e.g. 2002 in this study), (2) duration of epidemic **β** or how long does an epidemic persist, and (3) intensity of transmission γ or how significant do cases among people at risk happen and cumulate in consecutive weeks during an epidemic wave. A higher value of α, β, γ will indicate that a case studied is more likely to occur within a specific time interval with a longer duration and more intense transmission.

2.2.1 Probability of Occurrence Index (α)

α, the *probability of a disease* occurrence in certain week(s), is the proportion of one or more laboratory confirmed dengue cases occurred in an epidemic period (i.e.. numerator, *EW*) to the total number of weeks in that year (i.e.. denominator, *TW* = 52). The mathematical definition of α of a specific disease is expressed in Equation 1 below:

$$\alpha = \frac{EW}{TW} \qquad \text{Eqn. 1}$$

2.2.2 Duration of Epidemic Index (β)

To monitor the persistence of disease cases in a geographical area of interest, the *duration of an epidemic* (β) is described as the average number of weeks per epidemic wave as in Equation 2:

$$\beta = \frac{\sum_{i=1}^{EV} CW_i}{EV} \qquad \text{Eqn. 2}$$

where *CWi* represents the summation of persistent numbers of weeks for the ith epidemic wave, and *EV* is the total number of epidemic waves in the studying period. The duration index, reflecting the effectiveness of preventive and control strategies used during the early period of an epidemic, is very valuable for public health practitioners and administrators. A larger β value means fewer disease cases are eliminated.

2.2.3 Intensity of Transmission Index (γ)

To measure population-adjusted mean magnitude of dengue cases per wave, *intensity of transmission* equals to the mean number of case(s) occurred in successive weeks divided by number of people at risk in an epidemic wave. Therefore, the *intensity of transmission index* (γ) is formulated as in Equation.3:

$$\gamma = (\text{CaseNum/POP})/\text{EV} \qquad \text{Eqn. 3}$$

where *CaseNum* and *POP* represent respectively the cumulative total of confirmed dengue cases and population-at-risk, and *EV* as described above. The γ index is comparable to a measure of the intensity of cases in different areas considering both the duration and size of the population. A higher γ index implies that the occurrence of cases is temporally concentrated; whereas a lower value equates to a smaller epidemic wave, indicating that most cases are sporadic and temporally dispersed throughout the epidemic. The use of different temporal indices offers clues to key elements of an epidemic that are worth investigating.

2.3 Spatial risk index: Local Indicator of Spatial Autocorrelation

Other than temporally-defined epidemiological characteristics, the degree of spatial clustering among cases (i.e. spatial risk) (Odland 1988) is also needed to measure the degree of association between an interested temporal index and its specified location. Spatial autocorrelation, including global and local autocorrelation indices, are then introduced. A *Positive spatial autocorrelation* and *negative spatial autocorrelation* in the *Global indices* (Griffith 1987) respectively represent "clustering" of points based on tested variables (Anselin 1995) and inverse correlation, shown as "spatial outliers", between neighboring areas. A zero spatial autocorrelation is considered a random distribution rather than clustering or dispersal. *Local autocorrelation indices*, however, evaluate trends of clustering by comparing similarities and dissimilarities among neighboring locations. Therefore, further information about clustered loci is gained. In brief, the Local Indicator of Spatial Autocorrelation (LISA) can be regarded a spatial risk index to identify both significant spatial clusters and outliers (Anselin 1995). The definition of LISA index is given below:

$$I(i) = \frac{(X_i - \overline{X})}{\delta} \times \sum_{j=1}^{n} W_{ij} \times \frac{(X_j - \overline{X})}{\delta} \qquad \text{Eqn. 4}$$

where I(i) = the LISA index for region i; W_{ij} = the proximity of region i to region j; *Xi* = the value of the tested temporal index of region i; *Xj* = the value of the tested temporal index of region j; $\overline{X}$ = the average value of the tested temporal index; δ = the standard deviation of *Xi*; and *n* = the total number of regions to be evaluated.

The term *Wij* describes the proximity between region i and j, where a value of 1 means that region i is next to region j. The term $(X_i - \overline{X}) \times (X_j - \overline{X})$ describes the degree of similarity in a tested index within a designated area and its neighbors; from which each of temporal indices for the 2002 dengue epidemic in Kaohsiung and Fengshan was evaluated. A positive *I(i)* value of the tested LISA means that a certain region and its neighboring areas exhibit a clustering of homogenous areas and have a higher tendency of local spatial dependency. In contrast, a negative *I(i)* value, which shows the opposite trend between Xi and Xj (i.e. Xi = high, Xj = low or vice versa), implying a negative spatial dependency (i.e. the region is a spatial outlier in relation to its neighborhoods). The Monte Carlo significance test can be used to evaluate the statistical significance of spatial clusters and outliers (Anselin 1995). Risk areas can be classified by LISA index values into five distinct levels of epidemiological significance (Table 1).

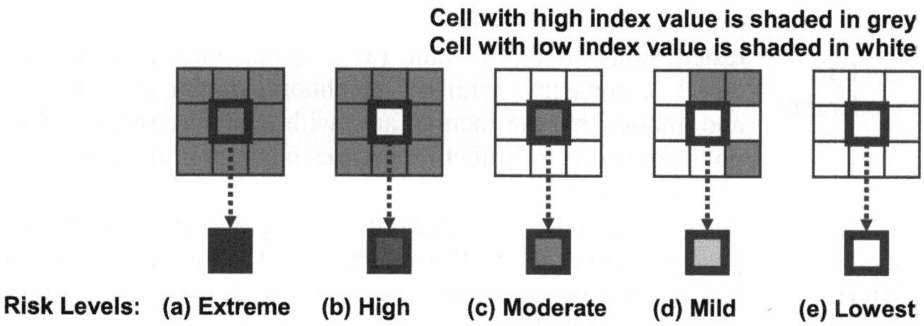

Fig. 2. Definitions of five spatial risk levels. Areas with a high value of the tested index are shown in grey and areas with a low index value are shaded in white. Among the five spatial risk distributions, (a) shows a spatial cluster of high values of the tested temporal index; (b) and (c) represent the spatial outliers; (d) shows random spatial distribution of the tested temporal index; and (e) displays a spatial cluster of low values of the tested temporal index.

Table 1. Five spatial risk levels with epidemiological significance

Spatial Risk Level	Epidemiological Significance
Level 1 (Extremely high)	Areas surrounded by neighboring areas having a statistically significant positive LISA index and a high tested temporal index value, shown in Figure 2(a), represent a spatial clustering of such an designated temporal index. This implies the presence of a severe epidemic spreading in that locality with extreme risk.
Level 2 (High)	Areas surrounded by neighboring areas with statistically significant negative LISA index, where the central area has lower tested temporal index value (as a spatial outlier) compared with it's neighbors, shown in Figure 2(b). This implies that a disease outbreak would soon occur in these high-risk areas, if effective control measures are not taken.
Level 3 (Moderate)	Areas surrounded by neighboring areas with statistically significant negative LISA index, where the central area has higher tested temporal index value (as a spatial outlier), opposite to Level 2, compared with it's neighbors, shown in Figure 2(c), and implies that the focused area with moderate risk could be a potential source of infection if cases begin appearing in the surrounding areas.
Level 4 (Mild)	Areas surrounded by neighbors having neither a significantly positive nor negative LISA index, shown in Figure 2(d). This refers to cases occurred sporadically in those areas but there is no significant spatial cluster or outlier.
Level 5 (Lowest)	Areas surrounded by neighbors having a statistically significant positive LISA index, but low for the tested all the three temporal index values, shown in Figure 2(e); this situation implies no outbreak occurs in those areas.

3 Results

3.1 Correlation among three temporally-defined epidemiological Indices

All three temporal indices (including probability of case occurrence α, duration of epidemic β, and intensity of transmission γ) show significant positive correlation with the "incidence rate" ($p < 0.001$) (Table 2). Among the temporally-defined characteristics, intensity is strongly correlated with duration (γ and β: 0.841), indicating that dengue cases in areas with a higher intensity are most likely to persist (i.e. last a longer duration). Yet, the correlation coefficient between α and γ at 0.425 is not significant enough, denoting that a higher occurrence of dengue cases does not always imply a higher intensity of transmission, which suggests difficulty in curbing the spread of the disease.

Table 2. Correlation matrix among incidence rate and the three temporally-defined characteristics in the 2002 Dengue Epidemic in Kaoshiung and Fengshan Cities in Southern Taiwan

	Probability of occurrence (α)	Duration of Epidemic (β)	Intensity of Transmission (γ)	Incidence Rate (IR)
Probability of occurrence (α)	1	0.625**	0.425 *	0.681**
Duration of Epidemic (β)	0.625**	1	0.841 *	0.636**
Intensity of Transmission (γ)	0.425 *	0.841 *	1	0.801**
Incidence Rate (IR)	0.681**	0.636**	0.801**	1

(*) means the value is high with a 95% statistical significance.
(**) means the value is high with a 99% statistical significance.

Nevertheless, sporadic cases likely to initiate an epidemic in certain geographic areas should be identified. Table 2 shows the importance of specifying spatial risk areas with different temporally-defined characteristics. Such information is valuable for public health officials to determine where dengue cases with a longer duration or a higher transmissible intensity but a lower incidence rate are located.

3.2 Spatial patterns analysis with temporally-defined epidemiological indices

This chapter illustrates a means to search for possible risk areas and point out likely areas with a longer duration of epidemic or more severe intensity of transmission. The procedures to examine each temporal index include the following: spatial mapping, spotting spatial risk areas, comparing their risk patterns, and finally, evaluating and filtering factors having the greatest influence on DHF epidemics in areas of different risk types.

3.2.1 Identifying epidemic risk areas

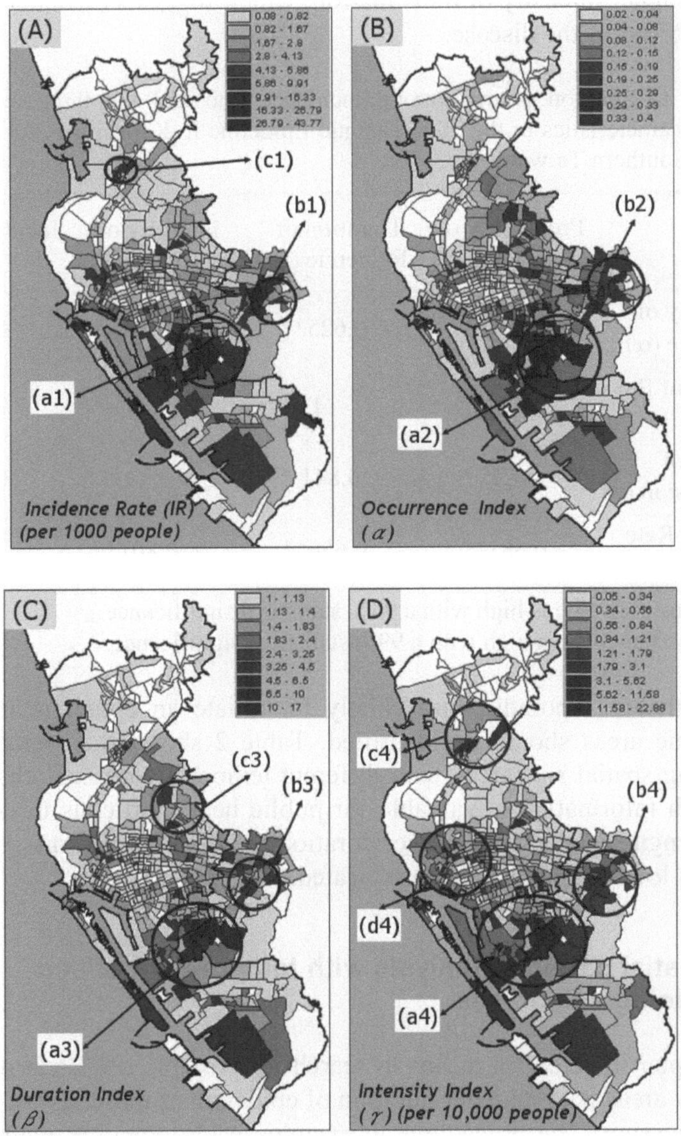

Fig. 3. Mapping the values of incidence and other three temporal indices (probability of occurrence, duration of epidemic and transmission of intensive) with observed clusters. The darker areas reflect a higher value of that indicated index. The locations of dengue clusters are shown as circled areas.

Figure 3 shows plots of the three temporal indices along with that of the incidence rate (IR). The spatial patterns on these plots are compared to identify possible risk areas (Mermel 2005). In these cases, darker areas in the figure reflect a higher value of the index. Risk areas with higher values in all three temporal indices (labeled as (a2)-(b2), (a3)-(b3), and (a4)-(b4) respectively in Figures 3B, 3C, 3D) are also recognized by IR (labeled as (a1)-(b1) in Figure 3A). However, areas with higher IR values (Figure 3A) are not detectable as areas with a longer duration of epidemic (e.g. c3 in Figure 3C) or those of a higher intensity of transmission (e.g. c4 and d4 in Figure 3D).

3.2.2 *Identifying spatially significant clusters and outliers*

"*Spatial outlier*" is another important spatial pattern for risk identification besides "spatial clusters," particularly at the beginning of an epidemic. To evaluate spatial association (including spatial clusters and outliers) with the three Li-based temporal indices, this study transforms LISA statistics into five different spatial risk levels with a 95% statistical significance (Table 1 and Figure 2) and displays their spatial distributions in Figure 4.

The LISA map for IR shows two statistically significant clusters (i.e. circles (a) and (b) in Figure 4A). Comparing this map of IR with that of the probability of occurrence (α), we can observe similarity in the clustering of cases in two locations (marked by circles (a) and (b) in Figure 4B). Similar clusters can also be found on LISA maps of the duration of epidemic (β) (i.e. circles (a) and (b) in Figure 4C) and the intensity of transmission (γ) (i.e. circles (a) and (b) in Figure 4D). Further comparison among the temporal characteristics in all areas marked (a) and (b) in Figures 4B, 4C, 4D suggests that areas with a higher α are larger in size and cover more Li than those with a higher β or γ (e.g. α: 14.35 km^2 vs. β: 8.74 km^2 and γ: 4.24 km^2 at circle (a); α: 4.25 km^2 vs. β: 0.47 km^2 and γ: 0.26 km2 at circle (b)). Fewer Li with a higher β or γ are found in circle (b) (Figures 4C and 4D), which implies somewhat effective control strategies. These areas in circle (b) characterized by having a shorter duration (0.7-1.5 weeks/wave) and a lower intensity of transmission (0.2-1.2 dengue cases/10,000 population-at-risk/wave) have had mosquito breeding sites reduced throughout.

Areas with 159 dengue cases can be identified using the IR map (9 Li covering about 1.05 square meters) although their values on duration and intensity are not particularly high (β = 1.92~2.01 weeks, γ = 1.18~1.26). Moreover, areas with moderate risks marked by circles (c) in Figures 4C and 4D have higher temporal index values than their surroundings (Central

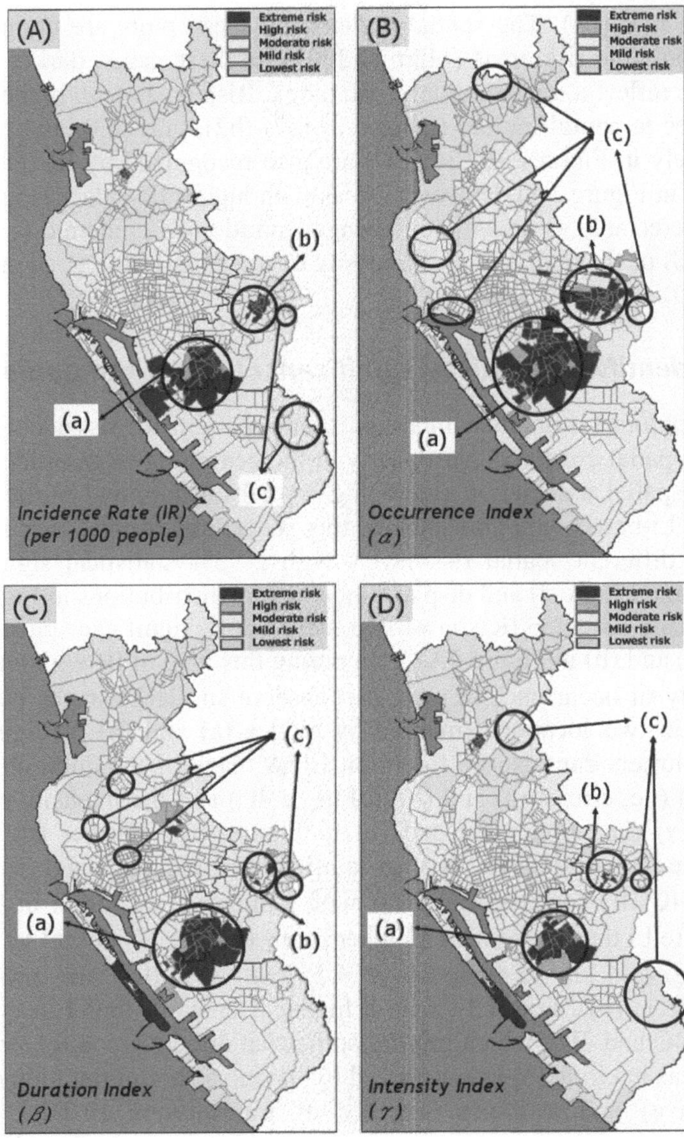

Fig. 4. Dengue significant risk maps of incidence and other three temporal indices to show spatial clusters and outliers. The Local Indicator of Spatial Autocorrelation (LISA) was adopted as the spatial risk index to identify both significant spatial clusters and outliers of the tested temporal in-dices. Monte Carlo significance test with $p<0.05$ was used to evaluate the statistical significance of spatial clusters and outliers.

vs. Surroundings: β=1.5~3.1 vs. 0.03~0.6 weeks, γ=1.04~2.01 vs. 0.01~0.57). These areas might experience further outbreaks if mosquito breeding sites were not eradicated. We can see that IR map alone (Figure 4A) without LISA map supplements (Figures 4B-4D) is not sufficient for distinguishing spatial risk patterns among epidemiologically significant areas.

4 Discussion

This chapter proposes a minimum data requirement and more straightforward statistical methods to capture major temporal characteristics along the dynamic epidemic process, and to suggest a workable procedure to intervene in endemic or hyper-endemic developing countries of dengue (Ali et al. 2003, Harrington et al. 2005, Morrison et al. 1998, Getis et al. 2003, Derek et al. 2004). The integrated spatial-temporal characteristics, unlike incidence rate alone, are useful in both nailing down risky areas and indicating places that might have virus transmissible to facilitate and cause another epidemic.

Temporal Epidemic patterns are complex. Although this chapter proposes three temporally-defined epidemiological characteristics (probability of occurrence, duration of epidemic, and intensity of transmission) to stratify the severity of an epidemic, it is difficult to compare these indices with each other in order of strength or severity measures. For example, which one of temporal risk patterns exhibited by, for example, a high intensity of transmission or a high duration of epidemic is more indicative? It is also apparent from the discussions above that none of the indices is able to stand alone. Clearly, further studies should focus on integrating three temporally-defined epidemiological indices into ONE single risk index. Furthermore, the effectiveness of this approach in identifying spatial-temporal epidemic risk patterns should be evaluated in depth by considering multi-year epidemic constructions, other diseases, and different locations.

In conclusion, the model proposed in this chapter can both quantitatively measure epidemiologically related temporal characteristics and qualitatively evaluate the effectiveness of control measures based on chronologically dynamic maps. Rather than relying on mapping cases or incidence rate solo at different periods, this study enables public health experts additional options: (1) to comprehensively identify risk areas and examine their dynamic spatial-temporal changes throughout an epidemic, (2) to provide broader perspectives with temporal risk characteristics other than epidemic curves, and (3) to measure risk levels along with the severity of epidemics after integrating spatial-temporal characteristics. Furthermore, monitoring spatial risk patterns over time described in this chapter can as well be applied to other infectious diseases, such as the West Nile encephalitis or Ebola hemorrhagic fever. Last but not least, the approach is potentially useful for controlling the fast spreading and highly pathogenic avian influenza virus (HPAI) by shortening the duration of the virus activity, and by minimizing its intensive transmission in epidemic sites; thus avoiding a possible pandemic (Mermel 2005).

References

[1] Ali M, Wagatsuma Y, Emch M, Breiman RF (2003) Use of geographic information system for defining spatial risk for dengue transmission in Bangladesh: Role for Aedes Albopictus in an urban outbreak. Am J Trop Med Hyg 69(6): 634 – 640

[2] Alpana B and Haja A (2001) Application of GIS in Modeling of Dengue Risk Based on Sociocultural Data: Case of Jalore, Rajasthan, India. Dengue Bulletin 25:92-102

[3] Anselin L (1995) Local indicators of spatial association – LISA. Geogr Anal 27 (2): 93–115

[4] Chao DY, Lin TH, Hwang KP, Huang JH, Liu CC, King CC (2004) 1998 Dengue hemorrhagic fever epidemic in Taiwan. Emerg Infect Dis 10(3):552-4

[5] Cockings S, Dunn CE, Bhopal RS, Walker DR (2004) Users' perspectives on epidemiological, GIS and point pattern approaches to analyzing environment and health data. Health Place 10(2):169-82

[6] Derek C, Rafael I, Norden H, Timothy E, Ananda N, Kumnuan U, Donald B (2004) Travelling waves in the occurrence of dengue hemorrhagic fever in Thailand. Nature 427(22): 344-347

[7] Dunn CE, Kingham SP, Rowlingson B, Bhopal RS, Cockings S, Foy CJ, et al. (2001) Analyzing spatially referenced public health data: a comparison of three methodological approaches. Health Place 7:1 -12

[8] Getis A, Morrison AC, Gray K, Scott TW (2003) Characteristics of the spatial pattern of the dengue vector, Aedes Aegypti, in Inquitos, Peru. Am J Trop Med Hyg 69(5): 495 – 505

[9] Griffith DA (1987) Spatial Autocorrelation: A Primer. Resource Publications in Geography (Washington: Association of American Geographers).

[10] Harrington LC, Scott TW, Lerdthusnee K, Coleman RC, Costero A, Clark GC, et al. (2005) Dispersal of the dengue vector Aedes Aegypti within and between rural communities. Am J Trop Med Hyg 72(2): 209 – 220

[11] Siqueira JB, Martelli CM, Maciel IJ, Oliveira RM, Ribeiro MG, Amorim FP, Moreira BC, Cardoso DD, Souza WV, Andrade AL. (2004) Household Survey Of Dengue Infection In Central Brazil: Spatial Point Pattern Analysis And Risk Factors Assessment, Am J Trop Med Hyg 71(5): 646–651

[12] Kuno G, Gubler DJ, Velez M, Oliver A (1985) Comparative sensitivity of three mosquito cell lines for isolation of dengue viruses. Bull World Health Organ 63(2):279-86

[13] Lai PC, Wong CM, Hedley AJ, Lo SV, Leung PY, Kong J et al. (2004) Understanding the spatial clustering of severe acute respiratory syndrome (SARS) in Hong Kong. Env Hlth Persp 112(15): 1560 – 6

[14] Lanciotti RS, Calisher CH, Gubler DJ, Chang GJ, Vorndam AV (1992) Rapid detection and typing of dengue viruses from clinical samples by using reverse transcriptase-polymerase chain reaction. J Clin Microbiol 30(3):545-51

[15] Mermel LA (2005) Pandemic avian influenza. Lancet Infect Dis 5(11):666-7

[16] Morrison AC, Getis A, Santiago M, Rigau-Perez J, Reiter P (1998) Exploratory space-time analysis of reported dengue cases during an outbreak in Florida, Puerto Rico, 1991-1992. Am J Trop Med Hyg 58: 287 – 298

[17] Odland J (1988) Spatial autocorrelation. Newbury Park, CA: Sage

[18] Shu PY, Chen LK, Chang SF, Yueh YY, Chow L, Chien LJ, et al. (2001) Antibody to the nonstructural protein NS1 of Japanese encephalitis virus: potential application of mAb-based indirect ELISA to differentiate infection from vaccination. Vaccine 19(13-14):1753-63

[19] Snow J (1936) On the mode of communication of cholera. The Commonwealth Fund. Oxford University Press, London

[20] Tran A, Deparis X, Dussart P, Morvan J, Rabarison P, Remy F, et al. (2004) Dengue spatial and temporal patterns, French Guiana, 2001. Emerg Infect Dis 10(4): 615 – 621

Public health, population health technologies, surveillance

A "Spatiotemporal Analysis of Heroin Addiction" System for Hong Kong

Phoebe Tak-ting Pang[a], Lee Phoebe [a], Wai-yan Leung [a], Shui-shan Lee[c], and Hui Lin[a,b]

[a] Department of Geography & Resource Management, Chinese University of Hong Kong, Hong Kong, China
[b] Institute of Space and Earth Information Science, Chinese University of Hong Kong, Hong Kong, China
[c] Centre of Emerging Infectious Disease, Chinese University of Hong Kong, Hong Kong, China

Abstract: Heroin addiction is a socio-behavioral and medical condition with significant public health implications. It is known that socioeconomic factors and population changes are important forces shaping the pattern of heroin addiction in a community. An understanding of these factors would be useful for the planning of effective intervention strategies. We applied the GIS technology in a study to display heroin addiction data over consecutive years in 18 districts of the territory of Hong Kong. Correlation was made between the proportions of heroin users in each district with the following variables: (a) changes in the local population size and structure; (b) economic attributes, in terms of employment and household income; (c) social factors, notably education levels; and (d) access to drug rehabilitation services, using methadone clinics as the surrogates. The data were obtained from available repository including the Census data, service statistics of methadone clinics, and Central Registry of Drug Abuse. The spatiotemporal framework has enabled analyses from different perspectives. We concluded that despite an overall decline in the number of heroin users in Hong Kong over the years, proportional increases were observed in selected districts. The association with poorer socioeconomic indices was obvious and the coverage of methadone clinics was found to vary from one district to another. The study offered a new means for examining conventional data, which may be useful in fostering public health policy development in the long run.

Keywords: GIS, heroin addiction, methadone clinics, spatial analysis

1 Introduction

Heroin addiction has always been a major social problem in Hong Kong. Although less than 1% of the total population was reportedly taking heroin in the territory, a figure much lower than those of 4-5% in nearby South-east Asian countries like Vietnam and Thailand (Department of Health, 2006), the potential threat of heroin addiction has continued to demand our attention. In regard to prevention and control measures, it is necessary to direct our attention to address socio-economic factors that may have triggered the abuse of heroin. Besides eradicating heroin-related problems, it would be beneficial to formulate measures to prevent people from heroin addiction in the first place.

Heroin is considered a "hard" drug to distinguish it from other psychotropic substances. The spatial distribution of heroin abusers is dictated largely by their addictive behaviors and access to instant treatment. The associated withdrawal symptoms when one is in need of heroin means that one is not expected to travel too far to have the problem fixed. In this connection, the Hong Kong government has established 20 methadone clinics to provide treatment on demand. We assume therefore that heroin addicts normally take heroin in their living regions and visit methadone clinics that are within their living districts. Spatiotemporal analysis of heroin addiction in Hong Kong would offer one means of describing the problem to inform effective policies on prevention and control.

The Narcotics Division of the Security Bureau of HKSAR has been, over the past 30 years, collecting drug abuse data through services like the hospitals and Methadone clinics. The information is open to the public through the Central Registry of Drug Abuse reports published twice a year. Heroin addiction patterns are often described with charts and tables but with little description of the geographic distribution. We piloted in this study the development of a system to support spatiotemporal analysis of heroin addiction in Hong Kong. The system involves not only the display of heroin addiction data by districts, but also offers a platform to visualize spatial pattern and temporal changes of heroin addiction, and to study the relationships between heroin abusers and their socio-economic characteristics. Data on methadone clinics are included in this system to allow users an option to investigate their accessibility.

2 System Design And Development

A GIS is used as a tool in this project to draw spatial relationship between the distribution of drug abusers and its correlation with districts and population characteristics, as well as the coverage of existing methadone clinics. The *Spatiotemporal Analysis of Heroin Addiction* system is designed to offer a more comprehensive view on the issue of heroin addiction in Hong Kong.

2.1 Data Source and Data Base Dictionary

A base map of Hong Kong was obtained from the Geography and Resource Management Department of the Chinese University of Hong Kong. Data on heroin drug abusers came from the Central Registry of Drug Abuse report. The population data for 2000 – 2004 were drawn from an annual report entitled "Population and household statistics analyzed by District Board district" published by the General Household Survey Section of the Census and Statistics Department. Location of methadone clinics and workload statistics were from the Department of Health and its website (www.dh.gov.hk).

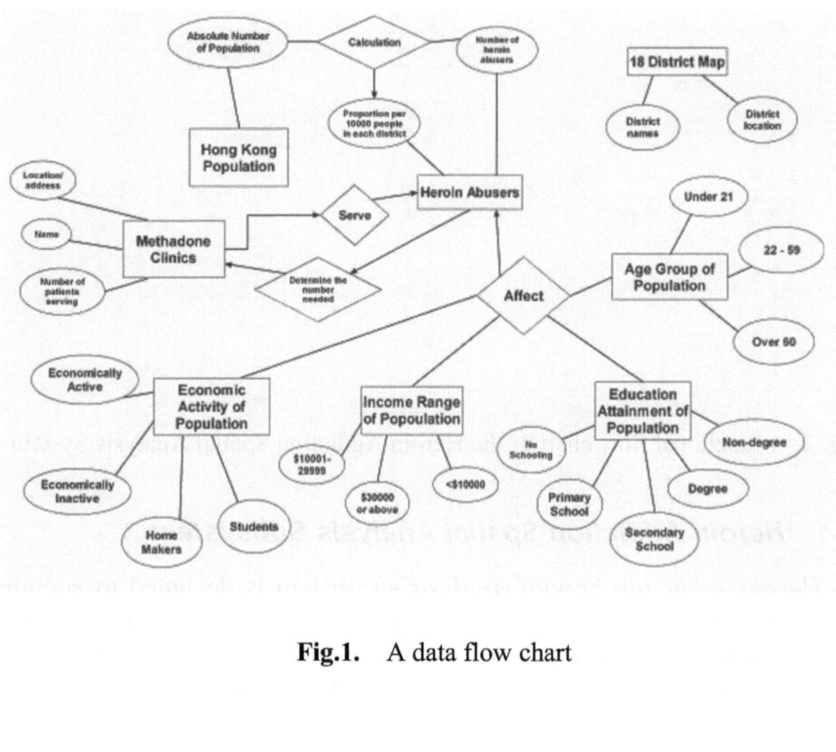

Fig.1. A data flow chart

To sustain a comprehensive analytical system for users, different types of data related to heroin addiction were prepared. These included data on population demographics, methadone clinic locations, and district level population statistics covering economic activities, income ranges, age groups and education attainment (Figure 1).

2.2 System framework

As a database system and also an analytical tool, the *Spatiotemporal Analysis of Heroin Addiction* system consists of two inter-related components: the *Heroin Addiction Spatial Analysis* and the *Methadone Clinic Spatial Analysis*. The existing database structure would support the organization of the datasets in a systematic way, making it easy for users to search data and carry out appropriate analyses with the help of analytical tools embedded in the System (Figure 2).

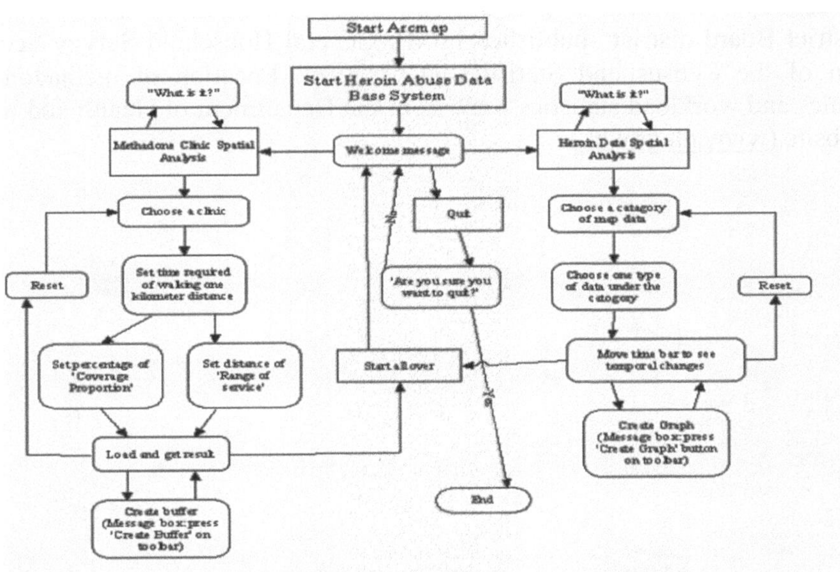

Fig. 2. A functional flow chart of the Heroin Addiction Spatial Analysis System

2.2.1 *Heroin Addiction Spatial Analysis Subsystem*

The *Heroin Addiction Spatial Analysis* subsystem is designed to support the search for relevant data and to analyze heroin addiction patterns and their relationships with population characteristics (i.e. income, education attainment, economic status, and age group). This tool emphasizes on temporal changes and specific statistical comparisons. To facilitate its use by non-geographers wishing to focus on the statistics alone, the system is designed to visualize spatial data in multiple representations. Heroin addiction statistics can be displayed over time, or as charts and tables with reference to their spatial locations at the District level. Users can observe heroin pattern changes over the years and the relationships with other population variables at the same time. A special toolbox is designed for creating graphs and buffers in support of further investigation.

2.2.2 Methadone Clinic Spatial Analysis Subsystem

The *Methadone Clinic Spatial Analysis* subsystem is an analytical tool for assessing the geographic range of service and coverage of heroin addicts of methadone clinics within a district. The system is designed on two assumptions: All heroin addicts are evenly distributed within the districts and the catchment area of methadone clinics appears circular in shape to yield the highest coverage extending from the clinics. The "Range of service" refers to the distance (or radius in this case) of the circular catchment area from each clinic while "coverage" of the heroin population refers to the proportion of heroin addicts that can be served within a district. Users can perform the calculation by either inputting the range of service to yield the coverage proportion or inputting the coverage proportion to get the maximum distance a clinic can serve.

Equation (1) is used to calculate the coverage of heroin addicts by inputting the range of service, while equation (2) is used to calculate the coverage of heroin addicts by inputting the coverage proportion.

$$\text{Range of service}^2 * \pi \, / \, \text{Area of the district} * \text{Total number of heroin addicts in that district} * 100\% \qquad \text{Eqn. (1)}$$

$$(\text{Coverage proportion}/\ 100 * \text{Area of the district}/\ 3.14)^{1/2} \qquad \text{Eqn. (2)}$$

The coverage of heroin addicts is further classified into different levels for evaluating the efficiency of methadone clinics. This system is designed for users who may wish to check the proportion of heroin addicts served by clinics within a specific district at different assumption levels of walking distances.

2.3 Software Requirement

The software required is the ArcMap GIS package, developed and marketed by ESRI. ArcMap supports various kinds of spatial analyses, visualization of spatial data and execution of different modeling functions. The Visual Basic Editor is also adopted for the programming of the system-interface.

3 Applications

3.1 Documentation

The *Spatiotemporal Analysis of Heroin Addiction 2006* system is equipped with two powerful subsystems. Users may click on the Analysis button to start either one of them. A Start dialog box with a welcome message will pop up to prompt users into selecting either the 'Hong Kong' option (representing the *Heroin Addiction Spatial Analysis* subsystem) or the 'Methadone Clinic' option (representing the *Methadone Clinic Spatial Analysis* subsystem). The 'What is this?' button next to the options offers an explanation of these choices (Figure 3).

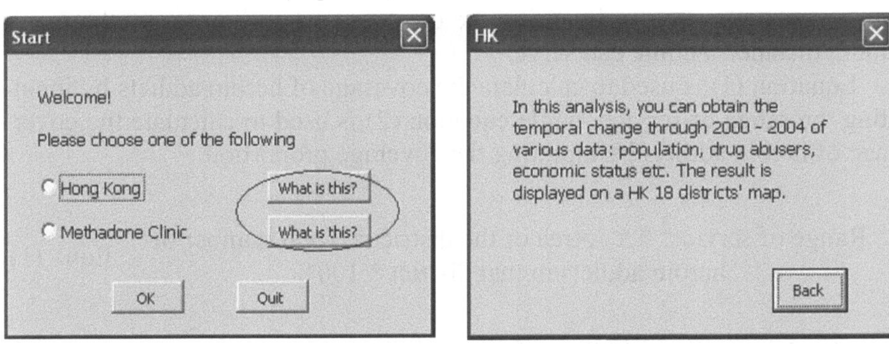

Fig. 3. The start dialogue box and the explanation message box

3.2 Applications of Heroin Addiction Spatial Analysis

The *Heroin Addiction Spatial Analysis* subsystem has two drop down lists. The first provides 6 categories of socio-economic factors related to heroin addiction, namely 'Number of heroin abusers', 'Number of population in each district', 'Education attainment', 'Income range', 'Economic activity' and 'Income group' (Figure 4). The second drop down list allows the choice of an attribute from the chosen category of data, and then the time bar will be activated. The legend of for the map display is shown at the bottom left portion of the dialog box.

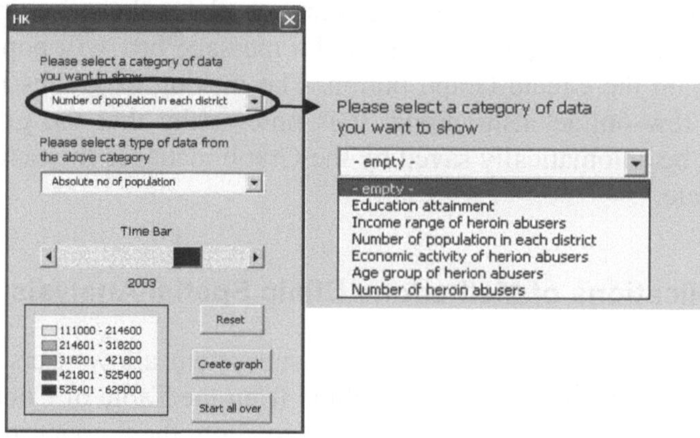

Fig. 4. The dialogue tool box of Heroin Addiction Spatial Analysis

In terms of map displays, users may visualize the chosen socio-economic data of a particular year represented as a choropleth map (Figure 5) and its progressive changes over time by dragging the time bar. The magnitudes of data in a choropleth map are represented in different shades, with darker tones indicating larger values.

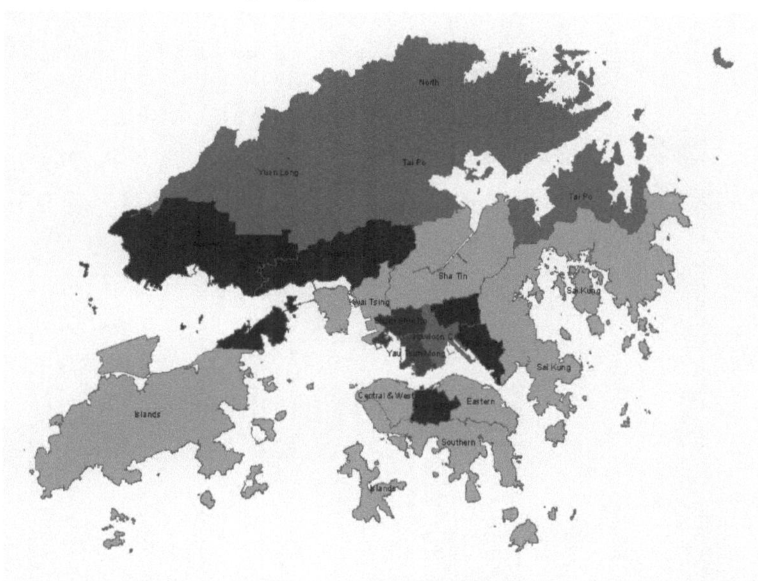

Fig. 5. A map showing variation of area-based data in different shades

Graphs/charts of the chosen data can be produced in place of area shadings by simply clicking on the Graph button which enables users to compare the observed results on both maps and graphs at the same time. To be more user-friendly, a Graph Wizard and a message box will pop-up when users click on the Create Graph button. The making of graphs is stepped through a few online instructions. It is noteworthy that the graphs produced will be automatically saved by the Graph manager for easy retrieval at a later date.

3.3 Applications of Methadone Clinic Spatial Analysis

The *Methadone Clinic Spatial Analysis* subsystem allow users to obtain basic information about a particular clinic from the Table of Content or all clinics by activating the layer number 20 to show them concurrently. The activated clinic layers will have the location of each clinics marked by a glass filled with a green-colored methadone. Details of a particular clinic such as the number of effective registration of methadone patients for each month in 2004 and particulars of that methadone clinic will be shown in a small window (Figure 6) by first clicking the Identify button on the tool bar and then placing the mouse cursor over the clinic on the map.

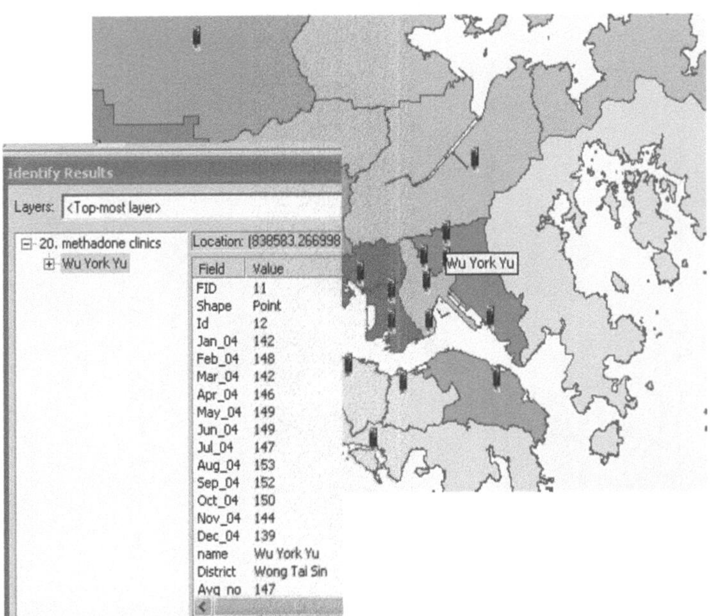

Fig. 6. Identify results of a clinic on the map

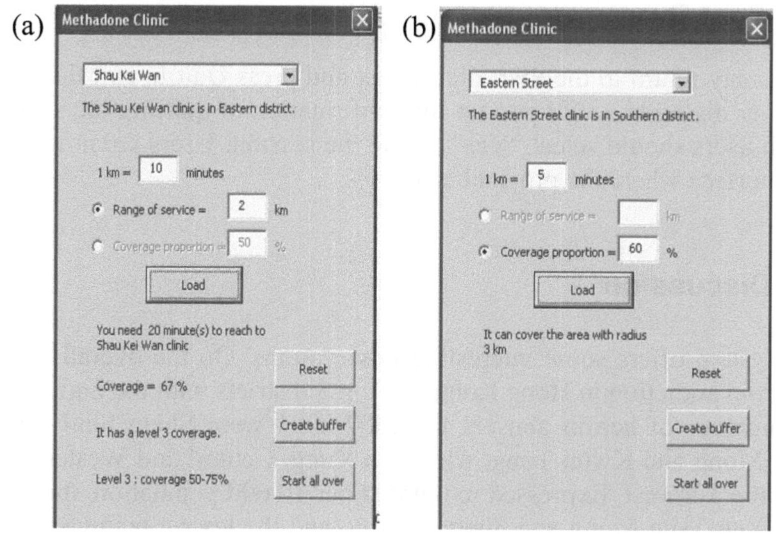

Fig. 7. Dialogue tool boxes for Methadone Clinic Spatial Analysis

With the *Methadone Clinic Spatial Analysis* subsystem, users can choose one methadone clinic from the drop down list in the Start dialogue box. Then, the district name of the selected clinic will appear at the bottom of the drop list. Users can key in the time equivalent in minutes of the walking speed of heroin addicts for a distance of one kilometer. The total required walking time to the range of services will be shown below the "Load" button (Figure 7).

There are two methods to assess the efficiency of clinics, which is either by determining the Range of Service (Figure 7a), or by determining the Coverage of heroin population (Figure 7b). For example, users can determine the proportion of coverage by adjusting the radius of the sphere of access for each clinic, or find out the maximum distance a clinic can serve for a certain proportion of the coverage. The input data and calculated results are shown in the space at the bottom half of the dialogue box. The level of efficiency is computed as the proportion of coverage in which a higher figure implies a clinic with greater efficiency where more heroin addicts in the population are served.

Users may proceed with another analysis on a different clinic by pressing the Reset button to make a new selection from the drop down list, followed by a new walking distance, range of service, or the cover proportion. The spatial coverage of each clinic can be displayed on a map having completed the calculation. The Create Buffer button can be used to invoke the buffer wizard to prompt users for values to draw buffers on a map.

3.4 Start Over

Users may return to the Start dialog box and press Quit to exit the system. Another dialog box will pop up for confirmation of quitting the system in which users should select “Yes” to end the session. Users are now back at the interface where the project began.

4 Discussion

The system offers some interesting observations. On the overall situation of heroin addiction in Hong Kong, the top 3 districts with the highest absolute number of heroin abusers for 2002-2004 were Sham Shui Po, Yau Tsim Mong and Kwun Tong, while Sai Kung, Central and Western Island were the 3 lowest. Expressed in number per 10,000 population, the highest were Yau Tsim Mong and Sham Shui Po and the lowest being Shatin and Sai Kung. There was a difference between the two comparisons due to variation in the district population. Sham Shui Po and Yau Tsim Mong were smaller districts with a low population and therefore registered a relatively large number of heroin abusers per 10,000 population. Kwun Tong on the contrary had a large absolute number of heroin abusers but the proportion was relatively small. In Shatin, the absolute number of heroin abusers was in fact not very high, but it was characterized by a high prevalence because of the small denominator.

Population structure was a determining factor in another observation. We found that the absolute number of heroin abusers in all districts decreased significantly but the decrease in the number of abusers per 10000 people was very small. For example, the number of heroin addicts in Tuen Mun decreased from 900 to 810 between 2002 and 2004, but the prevalence per 10,000 population remained the same. Despite an overall decline in the number heroin users in Hong Kong over the years, proportional increases were observed in selected districts, such as the Islands. It seems that the real situation of heroin addiction in Hong Kong has remained the same rather than improved.

A comparison of the population characteristics produces interesting results. Sham Shui Po and Sai Kung with approximately the same population size had the former registering the 2nd highest prevalence of heroin addicts while the latter showing the lowest number. Compared with Sham Shui Po, Sai Kung had a larger number of households of the high income groups, more people with a higher education, and a lower proportion of the elderly population. The number economically inactive population was also lower in Sai Kung. It can thus be inferred that population characteristics differ between districts irrespective of a higher or lower prevalence of heroin abusers expressed as a proportion of the population. Areas with a higher proportion of high income group, educated people, younger population and more economically active population tend to be associated with fewer heroin abusers.

Table 1. Coverage proportion of different clinics

Name of clinic	Level of coverage	District located
Aberdeen	3	Central and Western
Eastern Street	1	Southern
Shau Kei Wan	2	Eastern
Violet Peel	3	Wan Chai
Ho Man Tin	4	Yau Tsim Mong
Hung Hom	3	Kowloon City
Kwun Tong	3	Kwun Tong
Li Kee	3	Kowloon City
Ngan Tau Kok	4	Wong Tai Sin
Robert Black	4	Wong Tai Sin
Sham Shui Po	4	Sham Shui Po
Wu York Yu	4	Wong Tai Sin
Yau Ma Tei	4	Yau Tsim Mong
Cheung Chau	1	Island
Lady Trench	1	Tsuen Wan
Sha Tin	1	Sha Tin
Shek Wu Hui	1	North
Tai Po	1	Tai Po
Tuen Mun	1	Tuen Mun
Yuen Long	1	Yuen Long

* Level 1: 0-25%; Level 2: 26-50%; Level 3: 51-75%; Level 4: over 75%

Source: Department of Health, 2006

The *Methadone Clinic Spatial Analysis* yields other interesting findings. We set a walking speed at 15 min/km and obtained a range of service of 1.5 km. We then made an assumption that an average person or a heroin addict was prepared to walk 22.5 minutes to access a service. Table 1 shows that with the range of service kept constant, clinics located in Kowloon had the highest average coverage proportion and level of coverage. The level of coverage of clinics in Hong Kong Island was lower while that in the New Territories was the lowest. Apparently districts in Kowloon are smaller in size and therefore the clinics can serve a larger proportion of the districts and hence more heroin addicts.

Obviously, with the same range of service, it would be difficult for clinics in the districts of New Territories to cover a large proportion of areas. It is also understandable that the New Territories have fewer clinics as the absolute number of heroin abusers is generally smaller, thus less demand for the service. In some districts like Yuen Long, which has a relatively higher number of heroin abusers than other parts of the New Territories, there may not be sufficient methadone services to meet the need. From the above analysis, we may conclude that there is a need to review the geographic distribution of Methadone clinics to suit the latest trends and needs of heroin abusers.

5 Advantages and Limitations

This system is a pilot project built on the spatial method to support analysis of heroin addiction problems. There is the advantage of effective utilization of available data. Both heroin addiction data and population data are integrated in one system such that the inter-relationships can be observed and investigated. Numerical numbers and tables are transformed into spatial data or maps to illustrate concurrently their variations across districts and the patterns of distribution. Users can have a board overview of heroin addiction in Hong Kong. Another advantage is that the system supports a variety of functions like graphing and buffering to enable users to perform multiple analyses at the same time.

A major limitation of the system is however the small amount of data available given time limit and administrative problems. Data on population characteristics are insufficient to support a more extensive analysis, including the predisposing factors of heroin addiction. For example, it would be desirable to include in the analysis such fields like gender and housing type of population. However, it is difficult with limited heroin addiction data to see the long term temporal change of heroin abuse and perform future projection. While a considerable amount of data have been collected under the Central Registry of Drug Abuse, there are restrictions in acquiring data not currently included in the regular publications.

Another limitation is about the user interface. As a searching and an analytical tool, the *Analysis* system is generally useful to users. But given technical imperfections, users must exit the system to create graphs and buffers and they cannot use toolbar functions like zoom in and zoom out when the Analysis System is on. It would be more convenient if the Analysis System and the toolbar functions can be both functional at the same time. Other limitations include a slower loading speed for the graphics version without which users may not know exactly which layers should be chosen in the Graph Wizard to create the graphs they have in mind.

Despite the technical limitations, the system is unquestionably a valuable tool to transform numerical data into spatial data, along with support for the application of spatial analysis on a socio-medical issue.

6 Potentials and Future Development

The *Spatiotemporal Analysis System of Heroin Addiction* system is only the beginning of a new platform for gathering heroin abuse data in a GIS system and presenting it in a more organized way. The pilot system has compiled only 5 years worth of data. The next step would be an extension of the system to gather and organize data of earlier periods from 1976 to the latest available data.

The trend of drug addiction is changing. Heroin addition is usually found among people of older ages. There are many new substances of abuse such as Ketamine, Triazolam, MDMA (Ecstasy), Cannabis and Methylamphetamine (Ice), the distributive patterns of which have also been changing. With the collation of data collected by the Central Registry of Drug Abuse, not only would heroin addiction data be researched, but that data of other substances of abuse be input into the system to form a central information system, so that the trends can be analyzed effectively.

Currently, GIS-based systems are not widely available in the territory of Hong Kong. In the long run, the system should be developed into a web-based service for easy accessibility by people who need it. The system should consist of both ordinary (like the number of drug abusers in each district, or the address and number of heroin abusers served by each Methadone clinic in tabular formats) and complicated or statistical data (such as correlations between age and type of drugs abused). A web-based system would be an ideal solution for a simple and user-friendly implementation. Sharing of drug addiction data not only can increase the value of the data but also the feedback and opinion of users from different sectors of the society can improve the system by bringing in new and different points of view on the issue.

This analysis of heroin addiction can in the future help public health officials open up another dimension for research, that is the geospatial correlation of diseases transmitted among drug abusers. Knowing that needle-sharing among heroin abusers is one of the ways of transmitting blood-borne infections, knowledge on the pattern of heroin abuse can naturally help track the transmission of infections.

Acknowledgements

We thank the Department of Health for the provision of data on methadone clinics, the Narcotic Division for data from the Central Registry of Drug Abuse, and the Hong Kong Census and Statistics Department for the population data.

References

[1] Department of Health (2006) *Department of Health*. Retrieved on 28 March 2006 at http://www.info.gov.hk/dh/index.htm
[2] Household Survey Section (1) (2000) *Population and household statistics analyzed by District Board district*. Hong Kong: Census and Statistics Department
[3] Household Survey Section (1) (2001) *Population and household statistics analyzed by District Board district*. Hong Kong: Census and Statistics Department
[4] Household Survey Section (1) (2002) *Population and household statistics analyzed by District Board district*. Hong Kong: Census and Statistics Department
[5] Household Survey Section (1) (2003) *Population and household statistics analyzed by District Board district*. Hong Kong: Census and Statistics Department
[6] Household Survey Section (1) (2004) *Population and household statistics analyzed by District Board district*. Hong Kong: Census and Statistics Department
[7] Narcotics Division of Security Bureau (2004) Central Registry of Drug Abuse reports 54th Report 1995 – 2004, Narcotics Division of Security Bureau

A Public Health Care Information System Using GIS and GPS: A Case Study of Shiggaon

Ashok Hanjagi[a], Priya Srihari[b], and A.S. Rayamane[c]

[a] Associate Professor, Department of Geography & Geoinformatics, Bangalore University, Bangalore- 56 India
[b] Guest Faculty, Department of Geography & Geoinformatics, Bangalore University, Bangalore-56 India
[c] Professor, Department of Geography & Geoinformatics, Bangalore University, Bangalore-56 India

Abstract: Health data maps and Geographic Information Systems (GIS) are significant resources for health planning and health services delivery, particularly at the local level. The ability to visualize the spatial distribution of health status determinants and indicators can be a powerful resource for mobilizing community action to improve the health of residents. Currently, health data maps and other GIS applications tend to be highly technical and specialized, and are therefore of limited use to community members and organizations providing community-based health services. Developing relevant, accessible, and usable GIS and health data maps for communities and local agencies is an important step towards enabling individuals and communities to improve their health and increase their control over it. The final map was prepared by overlaying all the layers generated. The spatial objects were digitized out of LISS and PAN merged data and topomap supplied by the NRSA and Survey of India respectively. Questionnaires were prepared to get the data needed from each hospital and house by field investigation. Finally, a map of Public Health Care Information System was created by interlinking all topographical features with attribute data of the town so as to keep this information for planning and development in days to come.

Keyword: public health care system, GIS, GPS

1 Introduction

Health and life expectancy generally improve as societies develop but the major causes of illnesses and deaths will also change. The mapping of towns and cities (showing health centers, pharmacies, health camps, etc.) and creation of databases using recent technologies have become more significant both as an academic discipline and as one of the foundations for practical decision-making (e.g. in governmental, administrative and other public health care organizations) towards finding solutions of health related problems. It is generally known that the general public does not know where hospitals are located, what facilities and the kinds of specialists are available? Such problems can be addressed effectively by mapping and creating relevant databases using a GIS.

The call for streamlining of health care operations as a means of achieving greater cost-efficiency is a positive sign in the development of a society. Community leaders searching for innovative methods for health care management have begun to recognize the power of a GIS in various management activities ranging from determining intervention strategies to formulating health care reforms. Through the geo-coding process, a GIS allows personal health data to be examined spatially so that patterns can be discerned. Furthermore, geo-referencing of personal health data can greatly enhance decisions made by public health officials. The tremendous potential of a GIS to benefit the health care industry is just now beginning to be realized. Both public and private sectors (including public health department, public health policy and research organizations, hospitals, medical centers, and health insurance organizations) are beginning to harness the data integration and spatial visualization power of a GIS.

A GIS plays a critical role in the decisions on where and when to intervene, improving the quality of care and accessibility of services, finding the most cost effective delivery modes, and protecting patient confidentiality while satisfying needs of the research community on data accessibility. The GIS technology has been used in public health care for epidemiologic studies, as in tracking the sources of diseases and its spread in the communities such that authorities can respond more effectively to outbreaks of diseases with appropriate intervention measures to the at-risk population. GIS Applications in public health include tracking of child immunizations, conducting health policy research, and establishing health service areas and districts. A GIS not only provides a way to move data from the project level to become a ubiquitous resource for an entire organization but also renders the visualization of clinical and administrative data as a spatial decision support tool.

As towns have grown larger as a result of urbanization and industrialization, the concentration of population has in turn led to complicated health issues. A proper understanding of the roots of health problems is essential. In this connection, an understanding of the spatial occurrences of various diseases in urban area is a must for rational planning and management.

Knowledge about characteristics of the population and the diseases can be derived from GIS and Remote Sensing technologies.

Many health-related problems and determinants in the North Karnataka state of southern India have been discussed in research papers. These papers also shed light on causes, differential incidence and socio-spatial consequences of ill health within the towns. The town of Shiggaon, a taluk headquarter, was selected for this research. Shiggaon is situated between latitudes 14°59'24"N and 15°00'02"N and longitudes 75°12'36"E and 75 °13'48"E (Figure 1). The total study area measures 20.74 sq. km. with an annual precipitation of about 350 cm. Precipitation occurs in two seasons, mainly between June and September as well as between October and November. The mean annual temperature ranges from 22.8°C to 27.8°C. The population of Shiggaon consists of Hindu (70%), Muslim (29%), and others (1%). Socio-economic, demographic, and standard of living data of Shiggaon's population were gathered by questionnaire surveys. Projections on human needs were also made.

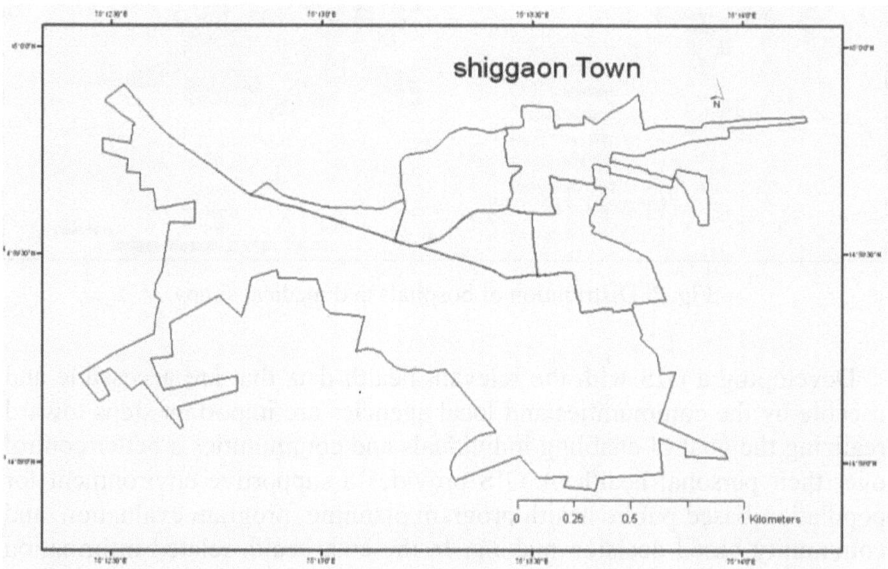

Fig. 1. A location map of Shiggaon town

The research aims at mapping Shiggaon to show the distribution of hospitals, clinics, and dispensaries, in addition to assembling a database on them. The public health care system concerns not only the health, well-being, and functioning of the population but also offers the ability to visualize spatial distribution (e.g., ward-wise) of health status indicators and other health-related information (Figure 2). Both GPS and ArcGIS software were used to demarcate the town's boundaries, locate hospitals and delimit disease vulnerable zones (Figure 3). Here, a GIS was used to collect, integrate and display population-based data concerning health events (including disease exposures, risk factors and socio-economic data).

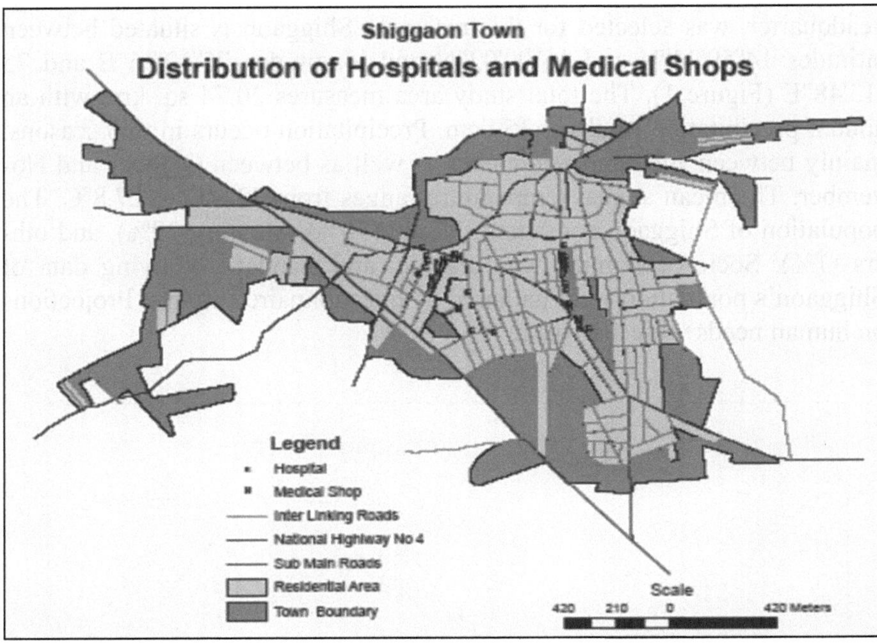

Fig. 2. Distribution of hospitals and medical shops

Developing a GIS with the relevant health data that are accessible and useable by the communities and local agencies are important steps toward realizing the goal of enabling individuals and communities a better control over their personal health. A GIS provides a supportive environment for population-based public health program planning, program evaluation, and community based decision-making. In the end, health-related information will be available for the public of Shiggaon located in the western parts of middle Karnataka (India) in the Haveri district.

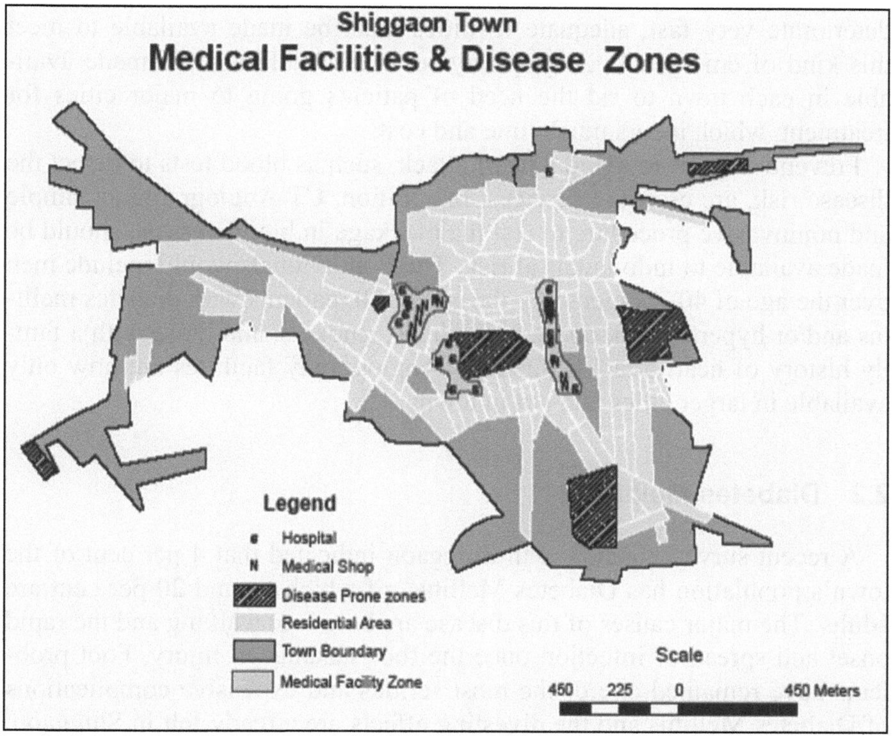

Fig. 3. Medical facilities and disease zones

2 Major Diseases in Shiggaon

2.1 Heart Related Diseases

WHO and various health organizations declared 2006 as the year of prevention. A major disease for prevention in Shiggaon is the cardiac disease and sudden cardiac death as these diseases have been spreading rapidly and reaching epidemic proportions. A 2005 survey of the 25-60 age-groups of Shiggaon revealed that half of the total death was from heart attack and heart-related problems. An even more startling fact is that one fourth of those with heart problems never reached hospitals to receive definitive treatment. If this was the fate of the population of a medium sized town, imagine the plight of people in the traffic chaos of Delhi, Mumbai, Kolkota, Chennai, Bangalore, and Hyderabad. Cardiologists have indicated that individuals with prior records of a heart attack have a higher risk of subsequent attacks. Given that the conditions of a heart attack patient can deteriorate very fast, adequate facilities must be made available to meet this kind of emergency. Angioplasty services should also be made available in each town to rid the need of patients going to major cities for treatment, which incurs much time and cost.

Preventive steps to avoid a heart attack, such as blood tests to detect the disease risk, are essential services. In addition, CT Angiography, a simple and noninvasive procedure to test the blockage in blood vessels, should be made available to individuals at risk. These individuals would include men over the age of 40, women over the age of 50, patients with diabetes mellitus and/or hypertension, obese individuals, smokers, and those with a family history of heart attacks. Such CT Angiography facilities are now only available in larger cities and not in towns.

2.2 Diabetes Mellitus

A recent survey conducted in Shiggaon indicated that 4 per cent of the town's population has Diabetes Mellitus, of which around 20 per cent are adults. The major causes of this disease are bare foot walking and the rapid onset and spread of infection once the foot sustains an injury. Foot problems have remained one of the most serious and expensive complications of Diabetes Mellitus and the divesting effects are already felt in Shiggaon. While advances in health care can help in early detection of the disease, educating the public on preventive measures are a must. For example, one should not walk barefoot indoor as well as outdoor and use correct footwear as prescribed by physicians. A daily inspection of the feet for blisters, wounds, bleeding, smell, or increased temperature is also recommended.

2.3 HIV/AIDS

Acquired Immune Deficiency Syndrome (AIDS) was first recognized in 1981. A large proportion of adults and children are infected daily. Because of high fatality and the lack of effective treatment or vaccine, the HIV/AIDS has emerged as one of the most serious health problems in India. In terms of income and wealth, AIDS has been found to associate with poverty and ignorance in Shiggaon. The percentage of HIV infected people living in the town should be higher than that reported officially as many are not aware that they are infected according to statements made by physicians. There were reportedly 55 confirmed AIDS cases in 2005 and about 20 percent HIV infected individuals in Shiggaon, according to sources from local hospitals.

Epidemiological studies have demonstrated that the major routes of HIV transmission are sexual intercourse, intravenous injections and transmissions from infected mothers to unborn fetuses through the placenta. Female Sex workers in Shiggaon have a significant level of HIV infection. A well-established National Highway No.4 which transects the town is known to be a major site of transmission of the virus where contact between sex workers and drivers is common. Increasingly, HIV is found to associate with sexually transmitted diseases (STDs) and tuberculosis, compounding an already alarming public health problem in Shiggaon. Unless serious preventive interventions are undertaken, there is a great potential for further acceleration in HIV prevalence.

The situation in Shiggaon can be summarized as follows. Firstly, HIV infection is rapidly spreading beyond the few high risk areas and exhibiting different epidemiological stages within the town. Secondly, interaction between HIV and sexually transmitted diseases and Tuberculosis is widely prevalent in the town, presenting an even more challenging public health problem. Finally, the co-relation between HIV and tuberculosis may result in a resurgence of tuberculosis.

2.4 Malaria

Malaria is a major health problem in Shiggaon where the carrier Anopheles mosquito breeds in stagnant pools of water due to poor sanitation and drainage. Other vector–borne diseases, such as dengue hemorrhagic and yellow fever, are associated with the need of households to store water in iron drums or earthenware containers which provide ideal breeding conditions for the Aedes Aegypti mosquitoes. The scarcity of clean water and lack of sanitation make diarrheal diseases a major health problem, while a variety of intestinal parasites, such as ascaris (roundworm) and trichuris (whipworm), are usually present. The crowded living condition also increases the risk of meningococcal meningitis, which leads to a high incidence of preventable infections in children such as measles, whooping cough and polio.

2.5 Tuberculosis

Tuberculosis is a contagious bacterial disease occurring in several forms in slum pockets of urban areas. It was once a common disease in Shiggaon but now found in migrant population returning from work in major cities of India. Major symptoms of Tuberculosis are fever, loss of appetite/weight, lack of energy, coughing, loss of color, irritability, and severe sweating at night. There were only a few cases of Tuberculosis diseases reported in 2005 by Government hospitals. However, studies have found that children under two and young women are more vulnerable than men or older people. There is no home treatment for Tuberculosis but the disease responds well to modern drugs. The key to avoid this potentially serious disease is frequent Tuberculosis screening.

3 Determinants of Public Health in Shiggaon

Awareness and behavior of the people play a vital role in disease prevention. The level of education of the mothers and the ability to read and comprehend health-related literature are keys to keeping family members healthy. Physical environment such as the residential location and its geographical surroundings has a direct influence on the health conditions of members of a family and the communities. Socio-economic conditions like income, education, religion are reflective of the conditions of health. All in all, air and water qualities, as well as the physical location of a settlement contribute collectively to the health of the people. Apart from these, biological and structural factors like age and gender or laws and norms play a vital role in the assessment of health care needs of the people.

The educational level of the residents of Shiggaon dictates their hygienic practices. About 30% of urbanities in the study area are illiterate and hardly resort to hospitals for their health problems. Water sources and sanitation facilities have an important influence on the health of urbanities. About 72% of households use piped drinking water, 15% use well water, and 13% use hand pumps. 50% of households either have a source of drinking water in their residence/yard or take less than 15 minutes round-trip to obtain drinking water. Only 60% of the town's population purifies their drinking water by some methods, including straining by cloth, filtering and boiling. The urbanities hardly use electronic appliances for water filtration and there is no proper treatment of drinking water by the local government.

With regard to sanitary facilities, the residents have hardly flush-type toilets and use either piped water or water from a bucket for flushing. Most of the urbanities have a pit drop toilet or latrine. However, close to 30% have no access to sanitary facilities, thus creating a lot of unhygienic conditions in the town.

4 An Integrated Approach to Health Care

An integrated approach is proposed here to bring together a range of health and other initiatives to produce an outcome in Shiggaon. Health improvement is seen only as a part of the integrated approach whose goal is total development of the community. This approach resembles the concept of a comprehensive Public Health Care (PHC) despite its potential need for building systematic linkages between physical improvements, social services, and resident participation.

Community participation is a key principle in an integrated approach. Environmental improvements may require some residents to give up parts of their plots or buildings to allow streets and drainage lines to be installed. Failure to consult local communities in advance often leads to a subsequent lack of cooperation and problems of ongoing maintenance of new infrastructures. Consent of the community, however, ensures that changes are more likely welcome, understood and longer-lasting which can also generate community spirit and promote local self-help. The potential of a grassroot-based integrated community development program is to be undertaken in Shiggaon. The local government and Non-Government Organizations (NGOs) are expected to mobilize local laborers and resources to develop an integrated program based on initiatives, physical upgrading, community development, and awareness of HIV/AIDS and other common diseases in the town. An effective implementation of the community development initiatives will require coordinated actions between the service providers and the intended recipients.

The complexity of poverty-related problems in the most deprived urban communities - including low levels of education, limited resources, and unfamiliarity with urban power structures - undermines the inability of the poor to mobilize resources for their health needs. Neither the NGOs nor private voluntary organizations (PVOs) can play a leading role in promoting community development. A public-private partnership between a city health authority and an NGO or PVO can provide the latter with a micro-scale perspective and framework for replication of successful projects. Public agencies will benefit by gaining a trusted mode of entry into the community and a means to deliver community-based PHC services with an emphasis on prevention as opposed to the more common facility-based curative care.

Figure 4 illustrates the processes to care for the health of urbanites in town. These processes involve a coordinated effort. Here, a technical team develops a respiratory health data model to facilitate identification and assessment of candidate data sources. The model attempts to describe the relationship between determinants and indicators of respiratory health. Criteria for evaluating candidate data sources were developed from a comprehensive metadata model developed for the project which included the following aspects of a data source: quality, completeness, relevance, ease of integration, potential for misinterpretation, and cost (if any).

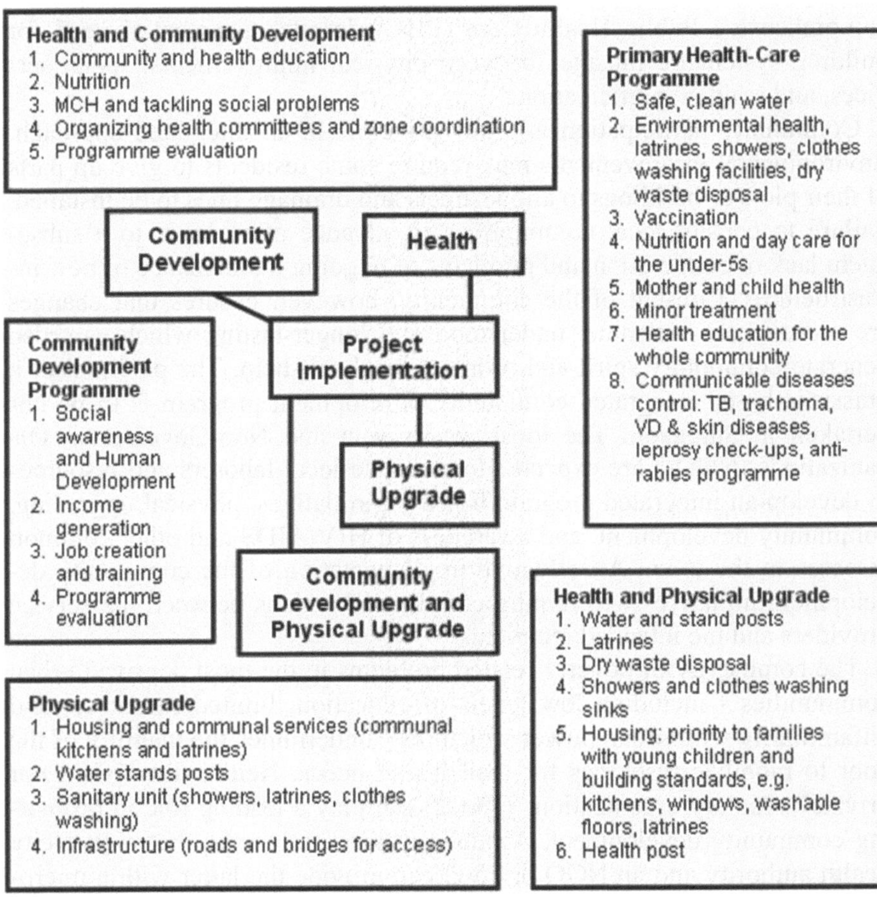

Fig. 4. The iterative process used to develop the health of urbanites in the town

5 Primary Health Care

The concentration of health resources and the relative proximity of hospitals and other medical facilities in cities have meant inconvenience to those living in towns as compared to those living in cities. Given that health-care services target preventive medicine, the underlying causes of ill health are unlikely to be addressed adequately. Primary Health Care (PHC) has emerged as the favored response to health care provisions in the towns.

At the heart of the PHC approach are the principles of equity in distribution, community involvement, a focus on prevention, use of appropriate technology, and a multi-sectoral approach that acknowledges the multiple aetiology of health problems. Major constituents of the PHC strategy include education about diseases and their controls, provision of safe water and basic sanitation, and measures to provide better maternal and child care, such as family planning, immunization against major infectious diseases, treatment of common diseases and injuries, and provision of essential drugs. Some have argued that this form of comprehensive PHC is idealistic and unattainable in a Third World city because priorities must be identified for reasons of practicality, cost and effective use of available resources given extensive health problems of varying severity.

6 Results and Discussions

Shiggaon has three distinctive seasons: (i) cold (winter) from October to March; (ii) hot (summer) from April to mid-June; and (iii) rain from mid-June to October. The town receives 85% of its rain from the southwest monsoon and 15% from the northwest monsoon. The general health conditions of its residents are from moderate to poor mainly because of poor sanitary conditions, drinking water, drainage etc. The proportion of preventive and curative clinical services is not up to the national mark and the amount of health services and programs are inadequate. The health and family welfare of Shiggaon are provided through a network of government or municipal hospitals and dispensaries and urban family welfare centers. Private hospitals, clinics and dispensaries also play a major role in providing these services to urbanities. While Shiggaon is not lacking of hospitals and clinics, these facilities are not well equipped to prevent occurrences of some diseases.

An important development took place when India adopted the National Health Policy in June 1981. This development may be viewed as an outcome of the declaration of health issues of the International Conference on Primary Health, jointly sponsored by the World Health Organization and UNICEF in Alma Ata in 1978 (World Health Organization and UNICEF, 1978). Although the National Health Policy places a major emphasis on ensuring primary health care to residents of all towns in India, it nevertheless identifies certain areas needing special attention in Shiggaon, as listed below.

1. Nutrition for all segments of the population
2. Immunization programs
3. Maternal and child health care
4. Prevention of food adulteration and maintenance of quality of drugs
5. Water supply and sanitation
6. Environmental protection
7. School health programs
8. Occupational health services, and
9. Prevention and control of locally endemic diseases

The healthcare system of Shiggaon is not reaching its intended population of 27,500 with another 50,000 living in its hinterlands. It also faces the risk of not having enough health workers to take care of the huge population. There is thus a need for a holistic approach to cater to the health care demands of its population. Anti-smoking measures, blood pressure and diabetes management advisories have been ignored by its residents who embrace a life of darkness and constant threat to healthy living. Despite efforts by government operated hospitals and private medical establishments in the town, there is little awareness on preventive health care. The fact is that people take their health for granted and are not bothered to obtain health insurances for preventive care. At the other end, water and garbage waterlogged soils are ideal for transmitting diseases such as hookworm. Likewise, pools of contaminated standing water provide breeding grounds for mosquitoes that convey enteric diseases like filariasis and malaria. Clearly, there is an urgent demand to improve the hygiene conditions through various measures, such as offering loans to households for upgrading shelters, granting supplies of cheap and easily available building materials, improving sanitary situations, and providing clean water supplies.

7 Conclusion

Population health is an emerging framework for assessing and evaluating the health status and health outcomes of a defined population. It is in many ways a superset of public health functions and goals. A GIS is an integral and essential component of a comprehensive population health information system. It is however a tool rather than an end in the practice of population health. The local government of Shiggaon has to take care of the general health welfare and prosperity of its residents by regulating the physical growth of the town and providing essential infrastructures for roads, water and sewerage. There is also the need to stimulate economic development, provide health education, and secure publicly accessible recreational opportunities. In doing so, the local government channels a variety of private interests toward decisions and policies about health and to protect public interests. The local government's efficiency, effectiveness and accountability must be improved in order to provide better health services to the people of Shiggaon. Nonetheless, active participation from the community is needed for the successful implementation of an integrated health program.

References

[1] Anselin L (1999) Interactive techniques and exploratory spatial data analysis. In: Geographical information systems: Principles, techniques, applications and management. Ed. PA Longley, MF Goodchild, DJ Maguire, DW Rhind. New York: Wiley. 253–66

[2] Batallha BHL and Parlatore AC (1977) Controle da qualidade da agua para consumo humano. Bases conceituaie eoperacionais (Control of water quality for human consumption). CETESB ediction. São Paulo: Brazil

[3] Briggs DJ (1992) Mapping environmental exposure. In: Geographical and environmental epidemiology: Methods for small-area studies. Ed. P Elliot, J Cuzick, D English, R Stern. Tokyo: Oxford University Press. 158–76

[4] Geographic Information Systems in Public Health, Third National Conference

[5] Lakoff G (1987) Women, fire, and dangerous things: What categories reveal about the mind. Chicago: University of Chicago Press

[6] Longley PA, Goodchild MF, Maguire DJ, Rhind DW. (1999) Geographical information systems: Principles, techniques, applications and management. New York: Wiley

[7] Loslier L (1995) Geographical information systems (GIS) from a health perspective. In: GIS for health and environment. Org. by P Wijeyaratne. International Development Research Centre, Ottawa

[8] Wartenberg D (1992) Screening for lead exposure using Geographic Information System. Environmental Research 59:310–17

GIS and Health Information Provision in Post-Tsunami Nanggroe Aceh Darussalam

Paul Harris[a] and Dylan Shaw[b]

[a]NGIS Australia & UN Information Management Service, Banda Aceh, Indonesia
[b]UN Information Management Service, Banda Aceh, Indonesia

Abstract: The large earthquake and subsequent tsunami that hit the coast of Nanggroe Aceh Darussalam (NAD), Indonesia in December 2004, was the cause of one of the largest natural disasters in recorded history. The health impact on coastal Acehnese community life was both immediate, with approximately 175,000 people either killed or missing, and also prolonged, with many health facilities destroyed and access to health professionals still severely restricted in many regions.

Providing medical services and supplies to those in need, rebuilding health facilities and increasing the capacity of the NAD Government to provide ongoing medical services has been a daunting logistic and administrative task, requiring timely and accurate information.

The information available to coordinate health recovery programs (much of which is spatial) has come from a variety of sources and been of variable accuracy and reliability. The United Nations Information Management Service (UNIMS) has worked with Provincial Government Departments, NGOs and other UN agencies for the last 14 months in Aceh, coordinating and validating much of this information and providing specialized advice and expertise to the humanitarian aid community to support planning and operational activities.

This chapter describes the UNIMS experience, focusing on how it supported health related activities. It describes the many challenges of acquiring and disseminating information in an emergency situation and how spatial information and technology are now being used to support Health planning. Lessons learned over the past 16 months in Aceh can be applied to other emergency situations, such as an outbreak of avian flu or SARs.

Keywords: health, GIS, tsunami, Aceh, earthquake, emergency

1 Introduction

On December 26 2004, an earthquake measuring 8.9 struck 150 km off the coast of Aceh. 45 minutes later, a tsunami wave hit Aceh devastating a 800 km coastal strip. Over 175,000 people were killed or remain missing. An earthquake on March 28, 2005 added to the destruction and casualties on the islands of Nias and Simeulue and in southern Aceh. In total, over 600,000 people were left homeless.

The physical destruction of the tsunami and earthquake included:

- 150,000 houses damaged or destroyed
- over 2100 schools severely damaged or destroyed (approximately 50%)
- 3000 km road impassable
- 120 arterial bridges destroyed
- all major seaports destroyed or severely damaged
- 8 hospitals and 114 health clinics damaged or destroyed
- 64,000 hectares of agricultural land and 15,000 hectares of aquaculture severely damaged or destroyed

In addition, these events caused immense social and economic devastation to areas where the people were already poor and had been suffering many years of conflict. The events also sparked unprecedented world-wide support and the largest emergency and recovery effort that the world community has undertaken.

The response initially focused on clean-up and providing temporary shelter and food for those left homeless, and emergency medical support for those injured. Remarkably, there was no major outbreak of disease or hunger, as so often the case after such disasters. Recovery programs are now focused on building houses, rebuilding infrastructure and restoring livelihoods.

The recovery effort is beset by challenges of enormous complexity. These include:

- Land had to be cleared of millions of tons of debris and silt before it could be used again – either for agriculture or to build houses.
- Land ownership had to be established before houses could be rebuilt.
- Large areas of land are no longer suitable for housing because of the impact of the earthquake.
- The single road along the west coast was washed away, resulting in a logistical nightmare for moving supplies.
- Water, sewerage and other services have to be planned before houses are built to ensure communities are viable.

- Avian flu and polio have reached Aceh (BRR, 2005).

Responding to emergency health needs and rebuilding the health infrastructure has been a daunting task. In addition to the destruction of health facilities, many health professionals died as a result of the tsunami. Careful planning and reliable information is crucial to make effective use of the resources available and to meet immediate needs.

2 United Nations Information Management Service (Unims)

Within days, the United Nations had established a Humanitarian Information Centre (HIC) in Banda Aceh. This Centre initially acquired available data for the effected areas, including satellite imagery, Indonesian Government and public domain spatial data. This information was provided to any agency working on recovery activities free of charge. Later, the HIC was also involved in analyzing the information to determine the extent of damage and to support the planning and logistics of food and water supply and clean-up activities.

As the emergency phase moved to the recovery phase, the HIC became the UN Information Management Service (UNIMS), producing a wide variety of maps and data products, collating and evaluating information about recovery efforts and coordinating information dissemination. At its peak, UNIMS employed over 40 staff, both international and Indonesian nationals.

3 Health Information Activities

3.1 Disease and Hunger

The most immediate concern was the fear that there would be an outbreak of disease or deaths due to starvation, as is so often the case after a disaster. Information was urgently required on the location of survivors and the extent of injuries. Herein lay the first challenge. Many areas were inaccessible. There were conflicting reports from NGOs as to the extent of the damage. Information from different government agencies often was inconsistent. Working with the World Health Organization (WHO) the HIC Team was able to evaluate the different sources of information and quickly put together a picture of the where the most likely areas of need were. This information was used to prioritize and coordinate activities. This included food and water supply and the provision of field hospitals and mobile health clinics.

The result was that there was no major outbreak of disease nor many deaths caused by starvation, although malnutrition is recognized as a problem in some area.

3.2. Health Facilities Damage Assessment & Rebuilding Capacity

The field hospitals and mobile health clinics provided response to the immediate needs of the survivors but rebuilding of the destroyed and damaged health facilities was required to restore the capacity to deliver health services to pre-tsunami levels. An assessment of damaged health facilities was undertaken to determine which ones could become operational more quickly and to prioritize rebuilding (based on population and proximity to existing facilities). Spatial information was used to support this process.

Most health facilities have been restored to pre-tsunami levels. Low levels of public investment and years of conflict resulted in the health facilities being ill-prepared to deal with a major emergency. Planning for longer term capacity building is now taking place. Programs are also required to tackle mental health problems, which are more complex and long standing than physical injuries.

3.3. Logistics

Spatial Information was used to support the supply and delivery of food, water and medical supplies. Because the main road on the west coast and many of the ports were severely damaged, alternative routes had to be identified and used. Efficient use of transport resources was also required. The UN Joint Logistics Centre and the World Food Program Shipping Service supported such agencies as the International Federation of the Red Cross to effectively plan and implement their delivery of medical and building supplies.

3.4. Contaminated ground water, pollutants and sanitation

After the tsunami, many ground water reservoirs were contaminated by saline intrusion as well as by a range of secondary pollutants from factories, oil facilities, septic tanks and storage facilities. Sanitation at many of the temporary shelter locations was also sub-standard.

Considerable attention was given to these issues to minimize the risk of disease. Drinking water was trucked in by NGOs, where required. Spatial information was used to identify risk areas and areas of need and to develop management plans.

As housing construction accelerates, the associated water and sanitation infrastructure needs to be in place. This is hampered by the fact that often different NGOs and agencies are responsible for different aspects of a community's rebuilding. This lack of coordination has resulted in houses being built without appropriate drainage, septic or sewerage systems. Spatial information can play a role in coordinating this activity. The Spatial Information & Mapping Centre (SIMC) has recently been established at the Baden Rekonstruksi dan Rehabilitasi (BRR), the Indonesian agency responsible for coordinating recovery activities, to support this process. However, it requires a mandate and direction form the BRR to the contractors to enforce provision of planning and as-constructed data for this coordination to be effective. Building the use of spatial information into a business process requires an understanding of spatial systems that is currently lacking by program managers in Aceh. One of the challenges is to increase this understanding.

In developed countries, these integrated solutions typically rely on the use of the Internet and possibly Web Services. In Aceh, the communication infrastructure is not currently adequate to support such solutions. However, intranet applications are possible and being introduced.

3.5 Indicators

Spatial information systems are being used to measure and monitor a range of health indicators and to identify areas that require further assistance. The maps below (Figures 1 and 2) provide examples of these indicators (BRR and UNIMS, 2005).

The indicators are used to develop and prioritize programs such as measles immunization, where midwives are required, responding to maternal deaths and combating child malnutrition. This information is also used to develop integrated health programs and identify training needs.

From these maps progress regarding the construction of health facilities can be gauged and cross-referenced with the incidence of disease to prioritize need.

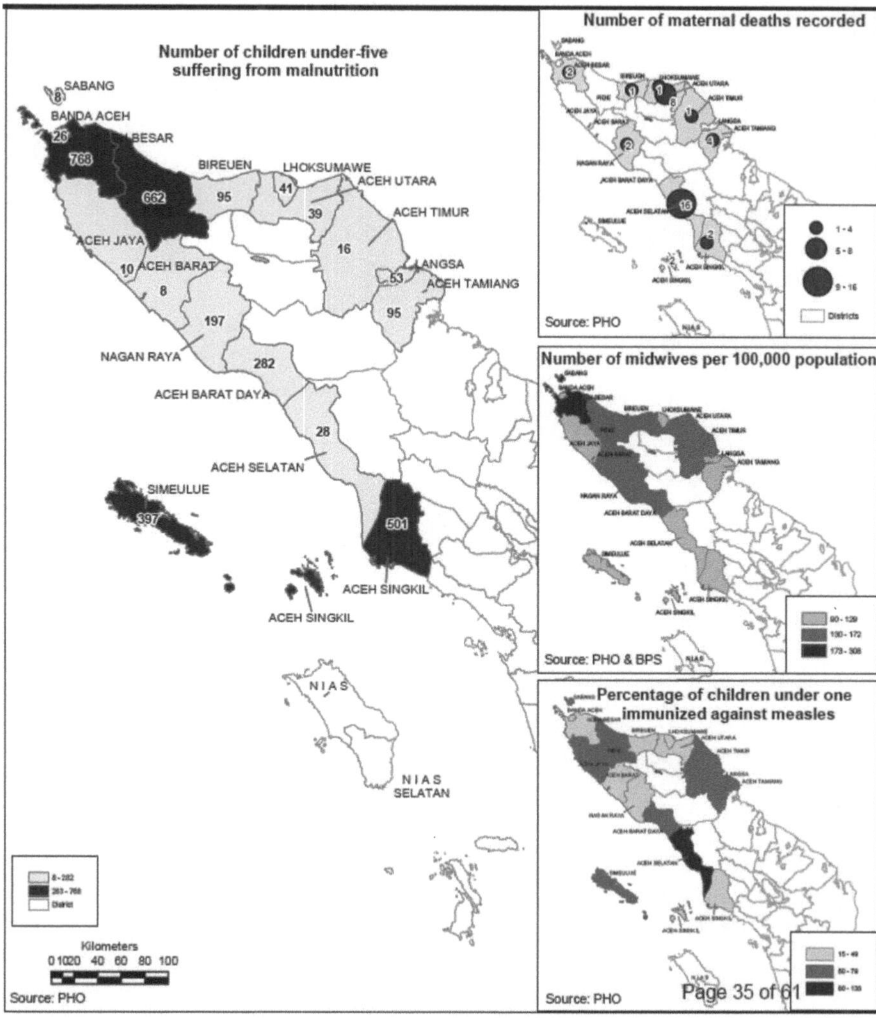

Fig. 1. Health related tsunami recovery indicators

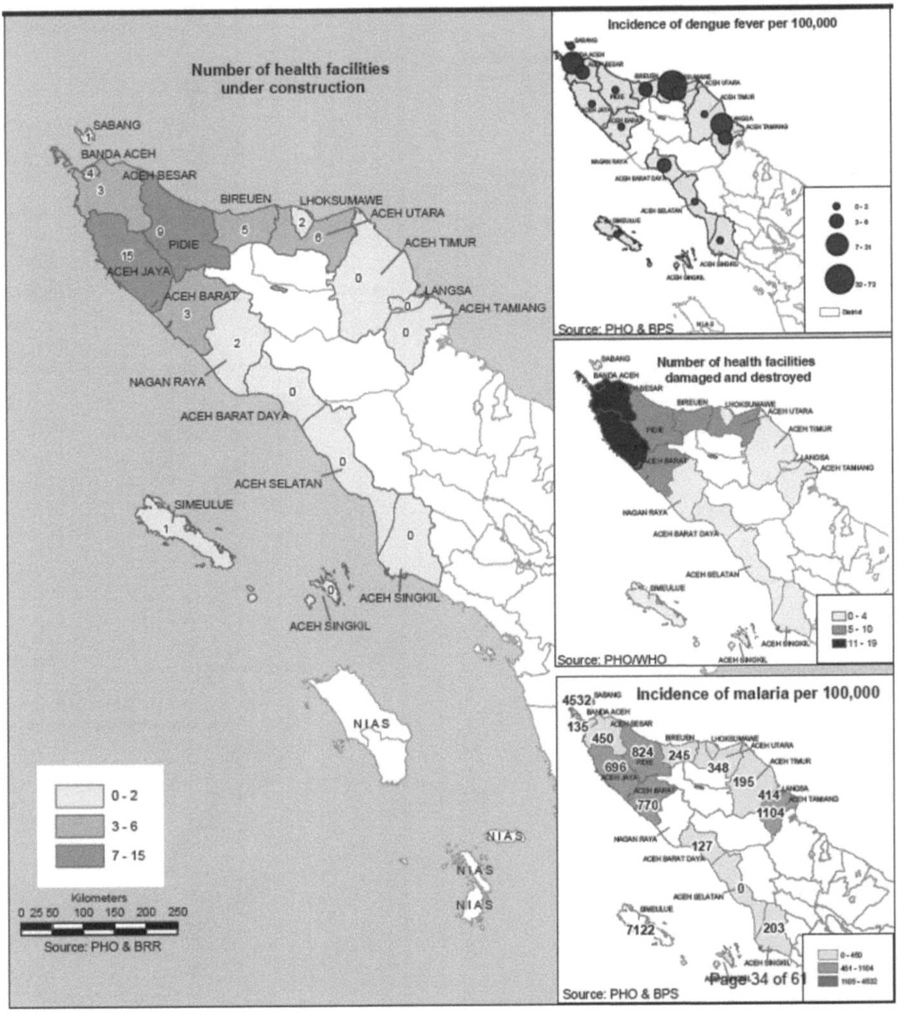

Fig. 2. Health facility and disease indicators

3.6. HealthMapper

In early 2005, the World Health Organization (WHO) provided technical support to the National Institute of Health Research and Development (NIHRD) and customized their HealthMapper application with data relevant to Aceh and tsunami affected areas (Figure 3). Information and analysis using HealthMapper was used to help plan much of the early response to the emergency and to better understand the impact of the tsunami on population health and to plan the longer term direction of rehabilitation efforts in the health sector. It was also planned to link HealthMapper to the proposed surveillance and early warning system. Staff at the Ministry for Health in Jakarta were trained to use the system and it continues to be used to monitor health related indicators.

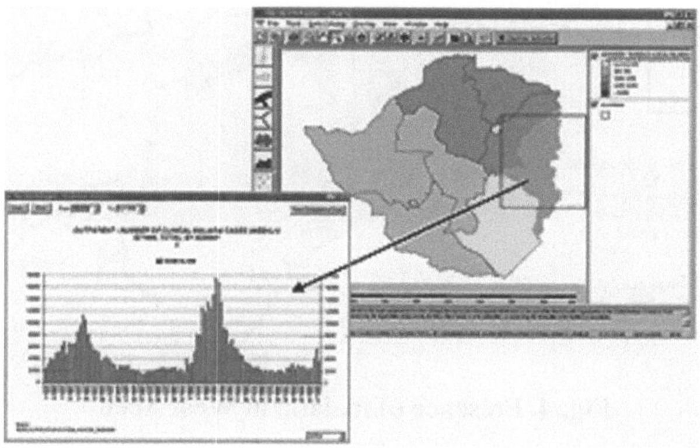

Fig. 3. HealthMapper interface: monitoring seasonal transmission patterns of malaria

However, there is little integration between this work and the planning activities at the BRR in Banda Aceh. While HealthMapper is a valuable tool that has been of benefit for a group of health-related professionals, this development is an example of the lack of coordination and collaboration that is occurring among agencies (both government and non-government) involved in recovery activities.

3.7 Malaria

There is anecdotal evidence that the incidence of malaria has increased since the tsunami. This is related to the extensive areas of stagnant water that have remained for many months close to populated areas. Early data indicated a high incidence of malaria along the west coast (Figure 4).

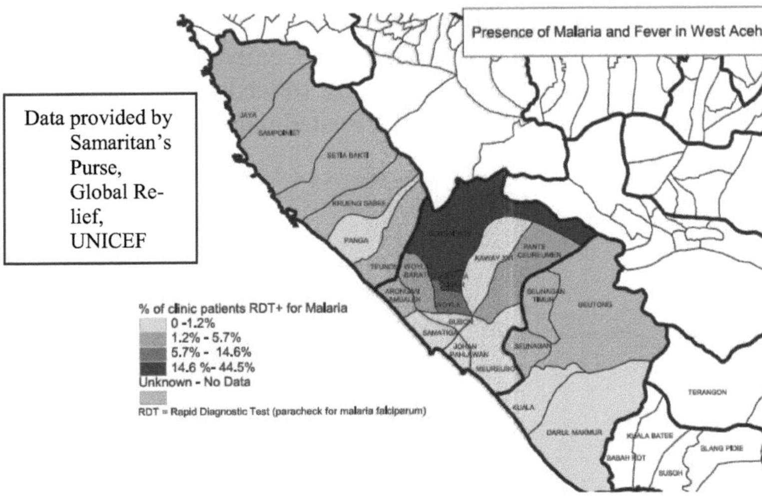

Fig. 4. Presence of malaria in West Aceh

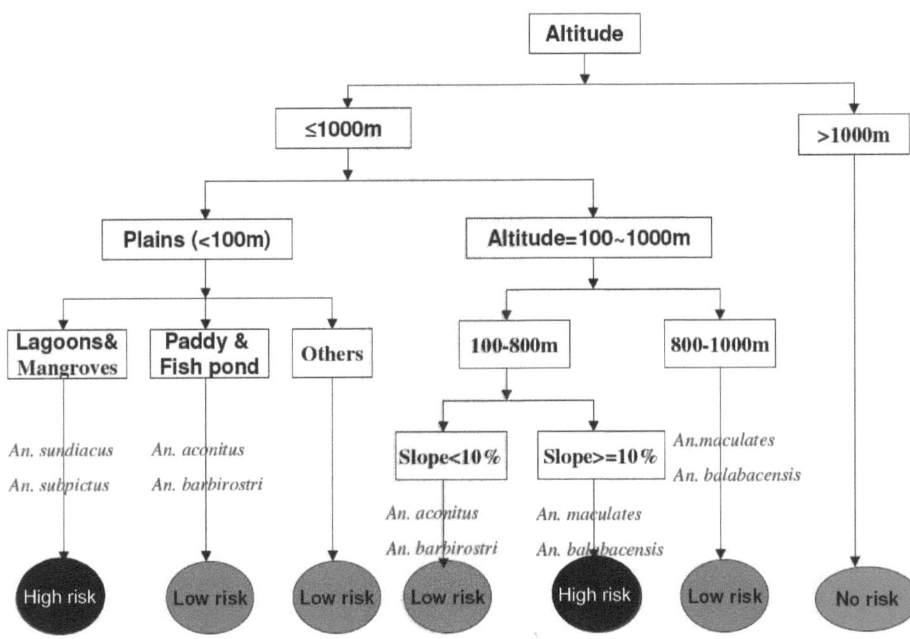

Fig. 5. Malaria stratification decision tree: malaria risk forecasting by GIS modeling

While the authors are unaware of any recent comprehensive analysis of the incidence of malaria in Aceh, GIS has been used in other parts of Indonesia to identify risk areas, develop time trends, and monitor malaria control and surveillance activities, to assess accessibility to health facilities and to forecast future potential malaria outbreaks. Dapeng Lou (2003) describes how GIS modeling is used to forecast malaria risks in Central Java, using a malaria stratification decision tree approach (Figure 5). His work demonstrated that GIS was an integral part of a malaria early warning system and that it could also be used to assess the effectiveness of surveillance and mosquito eradication measures.

4. Challenges & Lessons

4.1 Access to Information

The first challenge faced by spatial information officers in Aceh was to quickly find information about the area so that the emergency response could be planned and prioritized. The process of acquiring current information was severely hampered by the fact that, because of years of neglect and conflict, there was little up-to-date spatial data about Aceh and the Indonesian government was reluctant to make available what information there was. Satellite information and public domain data provided the first indication of the extent of the devastation and the likely areas where there would be the most casualties.

Many of the affected areas were inaccessible and information trickling back from these areas was frequently inaccurate, conflicting or incomplete. Information held by different levels of government was also often inconsistent. It was weeks before a complete picture of the damage could be put together.

Even when programs were developed, funded and implemented by Donor agencies, the information did not always flow back to the people or agencies who needed it. An example of this is the high resolution aerial photography that was flown in June 2005 and paid for by the Norwegian government. This imagery was in many cases the only current and reliable picture for villages that were destroyed by the tsunami and urgently required to support community planning and rebuilding activities. Yet it was not released until January 2006, largely because of decisions made by some sections of the Indonesian government to hold it.

The HIC worked to try and alleviate some of these problems by making what information it had available to any agency that required it. However, it did not have the mandate or resources to be in the field collecting information and was reliant on information given to it by other UN agencies, government departments and NGOs working in the field.

The lesson for government is to have a regular program for collecting information and a policy of data sharing and dissemination. The lesson for donors is to ensure that the work they are paying for will be available to all those who need it and not restricted by government agencies with their own agendas.

4.2 Coordination

The second and ongoing challenge is to coordinate the many projects occurring in Aceh, including gathering information. Lack of coordination has seen two or three times the number of required houses built in some villages and no houses built in others, houses built on land without access to drinking water, schools rebuilt where there are no longer children to attend them, property boundaries surveyed more than once; to name but a few examples. In short, lack of coordination has frequently led to either duplication of effort or no activity where some was required.

Information is crucial for ensuring that coordination can be effective. This starts with defining what information is required and then setting in process the actions to acquire the information, to a given standard. An agency needs to be given the responsibility of doing this. The role of the HIC and later UNIMS was to make available existing information, collating it where possible. While information gaps were identified, it did not have the mandate nor resources to initiate data collection programs. This was, in large, left to individual government agencies and a few NGOs, who undertook the task to inconsistent levels of accuracy and completeness.

As part of its coordination role, the BRR is in the position to initiate data acquisition programs. However, there is a lack of understanding of the importance of such programs and frequently information gathering is seen as an after-thought, rather than something that should be considered from the beginning of a project. The current Housing Program is an example, with many NGOs and contractors building houses without reference to each other or to other projects that are occurring in the same village, such as water supply and sanitation provision. It is recognized now that these projects should return as-constructed diagrams to the BRR but they typically are not referenced to any coordinate system and difficult to integrate with other data for the same area. If spatial information had been used to plan these projects from the beginning and the contractors provided with standard base maps, many of these problems would have been prevented. The SIMC is now working towards developing a pro-active approach to the use of spatial information at the BRR and among contractors and NGOs.

The lesson learned here is to make sure that spatial information is used in the planning process and that senior managers have a good understanding of what it can offer.

4.3 Collaboration

Collaboration is different from coordination. Collaboration involves the agreed working together of agencies to achieve mutually beneficial goals. While there was often the rhetoric of collaboration, in many cases the nature of the emergency, the independent culture of many NGOs, the design of projects and the culture of government agencies mitigated against it occurring.

This is not to say that there is no collaboration at all. Initially, there was extensive collaboration among Donor agencies and today collaboration is increasing at the grass-roots level.

The SIMC has established a GIS User Group, bringing together agencies and NGOs that have an interest in acquiring and using spatial information. This is resulting in collaboration among agencies involved in similar projects. Government agencies are also realizing that a cross-sector approach to information gathering is often required and there is a move to establish a government Spatial Information Coordination Group.

Collaboration only occurs when all parties realize that they will benefit from the collaborative efforts. A Forum needs to be in place to promote such an approach.

4.4 More than Maps

GIS is often seen as "making maps" and indeed during the initial response to the emergency the map making activities of the HIC provided immediate benefit for response planning. However, the challenge for spatial information professionals is to demonstrate how spatial information can provide more than maps and be a crucial part of the planning and management activities of an agency.

Spatial Information Managers must take the time to assess not only what information is required but how it can fit within the operational activities of recovery managers and then build systems to maximize it use. Resources need to be allocated for this activity.

4.5 Base Information

The recovery process in Aceh suffers from a lack of current base information. Much of the existing mapping is old and inaccurate. There is only partial coverage from satellite imagery and aerial photography. There was a delay in releasing the photography that existed.

The lack of base data contributed to many of the above problems, including the lack of coordination. Current maps, with a level of detail necessary for planning, remain unavailable for many areas of Aceh. This is severely impacting on what can be achieved in these areas.

The lesson for Donors here is that after a significant emergency, immediately put in place a program to acquire new high resolution orthorectified imagery (image maps) for the entire emergency area that can be used as a standard base for planning and monitoring.

4.6 Information Silos

The nature of information gathering after the tsunami led to many discrete databases being developed – databases about displaced persons, health facilities, damaged schools, housing needs, water logged land, destroyed bridges, land ownership, wells for drinking water, to name a few.

Rarely is there integration among these databases. Consequently, decisions are often made without knowing the full picture. GIS is an excellent means for integrating such databases. Browser-based solutions exist so that the information from disparate databases can be integrated and disseminated easily, to those who need it. Until recently, there has been little recognition of the need to do this.

The SIMC is working towards developing solutions to bring together these data silos so that all relevant information is available to program managers. The interface for such a solution can be seen in Figure 6. It enables all BRR staff to have access to information held by BRR about districts, villages and projects. Any number of separate databases can be linked to a unique mapping feature, providing a complete representation of the situation at the point or region.

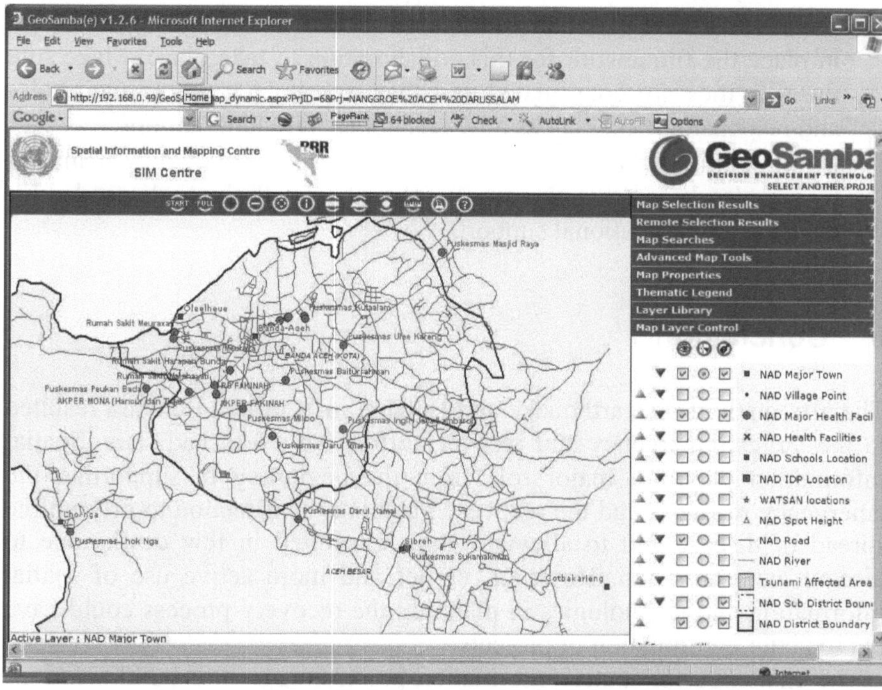

Fig. 6. Typical interface for intranet application used by BRR

5 Relevance for Other Emergencies

The magnitude and occurrence of the Aceh disaster was unforeseen. Emergency managers planning responses to an outbreak of SARs or avian flu are in a preferable situation. They know a disaster may occur and can plan for it. However, knowing that a disaster may occur does not always mean that the planning will be adequate, as demonstrated by the poor response by authorities to the aftermath of Hurricane Katrina in New Orleans.

Be prepared is obviously the message for governments and emergency managers. But what does being prepared entail?

- An accurate and up-to-date base dataset is part of being prepared. Know what information is available and know where the gaps are. Have in place a program to acquire information about the gaps and to capture post-emergency data, if relevant. Don't keep the data in silos – plan for data integration and sharing and have a system in place that will facilitate this.
- Make the data available to all who need it, free of charge.

- Ensure that there is collaboration among the relevant stakeholders. Put in place the framework for this collaboration to take place. Make sure the activities are coordinated and that stakeholders know their roles and responsibilities. Plan as a whole, not as individual groups.
- Build solutions that provide real benefit for the emergency manager, by seamlessly integrating spatial information with the tools needed for decision and operational support.

6. Conclusion

The aftermath of the earthquake and tsunami in Northern Sumatra resulted in the largest emergency and recovery effort the world had seen. Spatial information played a major role from the beginning in supporting the emergency response and the recovery planning. Swift action to prevent the spread of disease and to alleviate hunger resulted in few deaths due to disease and starvation. However, earlier and more active use of spatial information and technologies in planning the recovery process could have led to better coordination of projects.

The challenge for spatial information professionals is to build solutions that demonstrably prove the intrinsic value of spatial information. The challenge for health professionals is to ensure that they are effectively using the available information to plan and manage their activities.

References

[1] Dapeng Lou (2003) Geographic Information System: A Tool to Improve Decision Making on Malaria Surveillance and Control. Unpublished paper, Ministry of Health, Jakarta

[2] BRR (Baden Rekonstruksi dan Rehabilitasi) (2005) Aceh and Nias One Year After the Tsunami : The Recovery Effort and Way Forward. BRR, Banda Aceh, December 2005

[3] BRR and UNIMS (Baden Rekonstruksi dan Rehabilitasi; United Nations Information Management Service) (2005) Indicators Report: Final. BRR, Banda Aceh, December 2005

[4] WHO (World Health Organization) (2006) Public Health Mapping and GIS: The Health Mapper http://www.who.int/health_mapping/tools/healthmapper/en/index.html. Accessed on May 18, 2006

Estimating Population Size Using Spatial Analysis Methods

A. Pinto[a], V. Brown[b], K.W. Chan[c], I.F. Chavez[d], S. Chupraphawan[e], R.F. Grais[b], P.C. Lai[c], S.H. Mak[c], J.E. Rigby[e] and P. Singhasivanon[d]

[a] World Health Organization, Lyon, France
[b] Epicentre, Paris, France
[c] University of Hong Kong, Hong Kong
[d] Mahidol University, Bangkok, Thailand
[e] University of Sheffield, Sheffield, UK

Abstract: In population size is required within the first 24-72 hours to plan relief-related activities and target interventions. The estimation method should be easy to use by fieldworkers from various backgrounds, and minimize intrusion for the displaced population. Two methods have already been on trial: an adaptation of the Quadrat technique, and a newer T-Square technique. Here, we report the results of a field trial to test these alongside a newly adapted spatial interpolation approach. We compared the results with a population census of nine hamlets within the Tanowsri sub-district, Ratchaburi Province, Thailand. We mapped the study area to define the population for inclusion, as applications of this method would occur in closed settings. Before implementation, we simulated the spatial interpolation using geo-referenced positions of households in three hamlets. This procedure enabled us to establish some operational parameters to estimate population size, including the number of random points needed for the field test, the radius of the sample region and the dimension of the grid intersection on the interpolated surface.

Each method was tested over the same area. The interpolation method seems to produce accurate results at 30m-grid spacing (at 104% of the census) with a worst-case estimate at 124%. These results are comparable to those of the Quadrat (92% to 108%) and T-Square (80%). The methods are proven feasible to apply, with a high acceptability among local workers we trained. The interpolation method seemed the easiest to conduct. The results were tested statistically where possible, though this was an experimental setting, and further trial is recommended

Keyword: population estimation, crisis, GIS, spatial interpolation, field survey

1 Introduction

Population displacement during crisis conditions is frequent, and in most situations unpredictable. To plan relief-related activities and target interventions, an estimate of the population size (denominator) is required within the first 24-72 hours (Tellford et al., 1997). A good estimate of the population denominator also allows for the subsequent calculation of epidemiological indicators which permit the comparison between different areas and groups affected.

Non-Governmental Organizations (NGOs) often base their needs assessments on population estimates obtained from the modified Quadrat method (Grais et al, 2006). Recently, new methods such as the T-Square method have been tested (Grais et al, 2006). In recent crisis conditions, such as Darfur and Tsunami affected areas, Global Positioning Systems (GPS) and remotely sensed images were used to select geographical areas for sampling (Noji, 2005). Previous works conducted by Espie and others (Espie, 2000; Brown et al, 2001; Grais et al, 2006) have shown that the Quadrat and T-Square methods can provide reasonable population estimates.

Both the modified Quadrat and the T-Square methods are based on systematic random sampling within a spatially defined (closed) area. Sampling for the Quadrat method uses squares of equal sizes. The spatial distribution of units (shelters) present within the sampled squares is studied and their occupants counted to provide a basis for estimating the population of the entire site. Samples for the T-square method consist of pairs of units, using a nearest neighbor approach. An average occupancy of the units (shelters) is calculated from the pairs of units and the value is factored to obtain the count of individuals living in the whole site.

Traditional sampling methods are not designed for auto-correlated data and therefore the main purpose of traditional sampling is to avoid spatial autocorrelation and focus on obtaining averages from a sample of data. Spatial interpolation changes the emphasis to mapping the spatial distribution of populations. Using interpolation methods for a population in crisis may provide more robust estimates than traditional sampling. Further, interpolation methods may be less intrusive and quicker than other methods. Testing the feasibility and accuracy of spatial interpolation, never previously used for the rapid assessment of population in crisis conditions, is the main objective of this research.

Briefly, spatial interpolation is a method used to estimate an unknown value at a certain point using the known sampled values of surrounding points (ESRI, 2001). Interpolation methods have been employed mostly for continuous data, such as rainfall and soil chemistry, and incorporate spatial autocorrelation i.e. that closer values have a higher probability to be more similar than values which are further apart (Cromley and Mclafferty, 2002). The method of interpolation known as ordinary kriging is applied here.

2 Methods

2.1 Study Site

The study area is located in the Tanowsri subdistrict of the Suanphung district in the province of Ratchaburi in Thailand. The province of Ratchaburi is approximately 100 kilometers west of the capital city of Bangkok. It comprises nine districts with a total population of 791,217 in 2005. A long mountainous range surrounds the western side of the study area, while plains and occasional plateaus mainly occupy the eastern side. Altitude surrounding the area varies between 300 to 900 meters above sea level.

From the East, seventy-eight kilometers into Ratchaburi, is the district of Suanphung, which covers an area of 2,545 square kilometers. The district has one community hospital and 13 health centers. Suanphung has seven subdistricts with 8,254 households and a total population of 33,972. The Tanowsri subdistrict is within 8 kilometers of the Thai-Myanmar border. The majority of residents are of Karem ethnicity most having originally migrated from Myanmar. The population is now relatively stable with very limited in or out migration. Settlements are constrained by the topography: fairly steep-sided, narrow valleys. This is an important factor influencing the spatial dispersion of the households.

2.2 Population Estimation Methods

Two weeks before the field work, an exhaustive census was carried out in the nine hamlets of the study site. All household members were recorded by two age groups: under 5 years-old and >= 5 years-old. Each household was marked with a yellow number for easy recognition by the field teams. The definition of a member of household included individuals who slept there the night before.

Although the area had been mapped for the census, the first stage of the field work was to redefine the boundaries of each hamlet to reduce heterogeneity across the area. The field teams used Global Positioning System (GPS) for this mapping. Some isolated, scattered households were excluded. The resulting survey population (Table 1) is thus smaller than the census population. Hamlet C shows a big difference between the two populations because most of the area was inaccessible due to flooding at the time of this revised boundary mapping exercise. For this reason the area was subsequently excluded from the field test. Hamlet G was also excluded because of its small size and geographic separation.

Table 1. Hamlets in the Tanowsri subdistrict of Thailand

Hamlet	Survey population Used as gold standard
D – Pong Haeng	289
X – Huay Muang	524
Z – Borwee	781
A – Huay Namnak	878
B – Wangko	205
C – Huay Pak	145 (Not included)
E – Huay Krawan	258
F – Nong Tadang	236
G – Phurakham	Not included

2.3 Interpolation

Interpolation usually requires a dataset of measurements at randomly sampled points across the study area. Clearly for a population estimate, a point measurement is inappropriate. The team therefore decided to generate the measurement in terms of the population density of a circle surrounding the point. Two parameters: the radius of the circle, and the number of points to be sampled, were estimated from computer simulations of the study area. The radius of the sampling circle was established as 25 meters, and the sampling set at one point for every $150\text{x}150\text{m}^2$ of area. The interpolation method was tested only in three Hamlets: A, X and Z. The main reason for limiting the field test to these three hamlets was the difficulty in generating random points in the remaining hamlets given their small areas. The number of random points needed for Hamlets A, X, and Z (the three largest areas) is shown in Table 2:

Table 2. Number of random points per study area

Hamlet/polygon	Area in m^2	Number of random points
A	382411	18
X	205759	9(+1)*
Z	179983	8(+2)*

* It was noted that the software requires a minimum of 10 random points to interpolate a surface.

The interpolation method was conducted twice in each of the three hamlets using different sets of random points. The points were computer-generated and downloaded into a GPS so they could be located in the field.

Once a random point was reached in the field, a stake attached to the end of a 25m rope was driven into the ground to facilitate the construction of the circle circumference centered on that random point. In the first instance, a visual inspection determined whether there were any households in the circle or not – the use of random points meant it was quite possible to have an 'empty' circle. Otherwise, four points on the circumference of the circle were established by measuring with the rope and marking with a bamboo stake. If it was still not possible to determine whether a household was fully in the circle, the rope was extended out to the household. Clearly there were households which lay on the circumference of the circle, necessitating a method of deciding whether they should be included in the sample measurement or not. The team decided that all households inside the circumference and touching the circumference should be counted; recording which household was totally inside and which crossing the circumference. The household data was extracted from the population census in reference to the yellow identifying number on the house.

2.4 Quadrat

We applied the Quadrat method as described in the Epicentre MSF field test in Mozambique which suggested that using 15 quadrats (625 m^2) yielded an acceptable estimate of population (Grais et al., 2006). A total of 38 points was randomly and proportionately distributed over the surface areas of the seven disjoint hamlets (Table 3). The points were generated using the Excel application EPOP v.1.23 (Epicentre 2005). Each quadrat had a side length of 25m giving a total area of $625m^2$.

Table 3. Number of random points for the Quadrat method

Hamlet/polygon	Area in m^2	Number of random points

A	382411	11
B	54706	3
D	69324	3
E	54741	3
F	34689	3
X	205759	8
Z	179983	7
	Total =	38

Again, a GPS was used to locate each point, the quadrat constructed, and households recording, following the published protocol as to whether households on the boundary should be included or not.

Population estimates were calculated using the Poisson and negative binomial models. A Poisson model is commonly used for such estimation but its applicability is limited to randomly distributed units. A Poisson model assumes that: 1) each unit has an equal probability of hosting an individual; 2) the occurrence of an individual in a sampling unit does not influence its occupancy by another (i.e. independent events); and 3) the mean (μ) is equal to the variance σ^2. Below are 3 basic patterns and their variance-to-mean (VMR) relationship, where σ^2 = variance and μ = mean:

1. Random pattern: $\sigma^2 = \mu$;
2. Clumped pattern (over dispersion): $\sigma^2 > \mu$
3. Uniform pattern (under dispersion): $\sigma^2 < \mu$

If such assumptions are not met, then a Poisson model is not applicable and appropriate distribution should be considered: negative binomial distribution for clumped patterns and the positive binomial distribution for uniform patterns. (Ludwig et al., 1988).

Using population data from all 38 quadrats, the average population per quadrat was calculated. The mean quadrat occupancy is given by dividing the total weighted frequencies with the total number of quadrats surveyed. The calculated mean (from either Poisson or Negative binomial procedures) serves as an estimator of population density, thus the estimated population size is given by:

$$\text{Estimated population} = \frac{\text{mean quadrat occupancy} * \text{total area (in m}^2\text{)}}{\text{quadrat size (in m}^2\text{)}} \qquad \text{Eqn. (1)}$$

2.5 T-square

We adopted the same T-Square methodology applied by the Epicentre in Mozambique in 2004 (Grais et al., 2006). The T-Square method was also tested in all seven hamlets selecting a total of 82 points as guided by the Mozambique study which suggested the selection of at least 50 randomly distributed points in the seven hamlets as showed in Table 4:

Table 4. Selection of random points for T-square method

Hamlet/polygon	Area in m^2	Number of random points
A	382411	26
B	54706	6
D	69324	6
E	54741	6
F	34689	6
X	205759	17
Z	179983	15
Total	=	82

For the T-square method, once a random point was reached in the field the team would measure the two distances as showed in Figure 1. A decameter was employed to measure short distances while the telemeter was used for distances over 20m. Where possible and in most cases, distances recorded by the telemeter were also counter-checked using the decameter or the GPS.

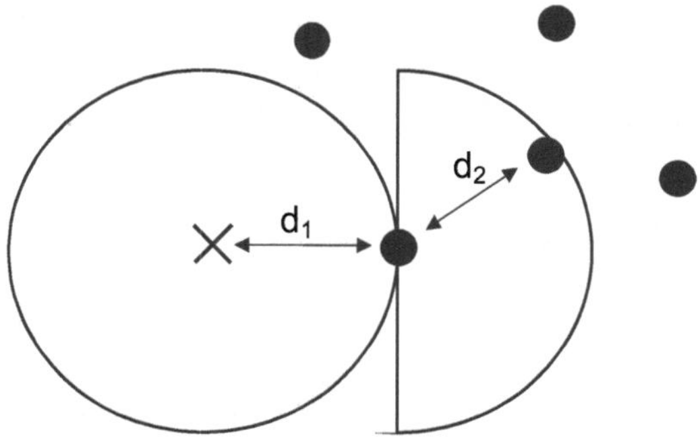

Fig. 1. Selecting households for the T-square method

The T-Square method allows for the calculation of the average unit occupancy of the number of shelters drawn at random in the site. The number of persons living in the shelters is counted and the average number of persons per shelter is computed. The total number of shelters in the site is obtained by dividing the total site surface with the average shelter occupancy. The total population for the site can be obtained by knowing the average surface occupied by a shelter. Depending upon results of the heterogeneity tests, the estimation can be presented with or without their confidence intervals.

2.6 Data analysis

Interpolation

We applied the ordinary Kriging interpolation method using the Geostatistical Analyst extension of ESRI's ArcGIS 9 software:

1. Population density for 25m buffer zone was computed.
2. Data were entered in an Excel file and thus related by house code to the GIS database.
3. An interpolation surface was generated using ordinary kriging on population density
4. 30m, 35m, 40m, 45m and 50m grids were generated for population estimation.
5. Grid points were used to extract the population estimated and results compared against true population census data

Quadrat and T-Square

For the analysis of Quadrat and T-square methods, EPOP software (Epicentre, Paris, France) was used. However, this was extended using a negative binomial approach (explained above).

2.7 Simulation Analysis

Because the field survey for interpolation was carried out only twice in each of the three selected hamlets, we used simulation analyses to extend the testing. In general, the estimation capacity of the interpolation method was tested using multiple sets of field and simulated data as summarized below.

Table 5. Number of random points for hamlet and field survey conducted

Hamlet/polygon	No. random points	No. Field surveys	No. simulations
A	18	2 sets	20 sets
X	9(+1)*	2 sets	10 sets
Z	8(+2)*	2 sets	10 sets

3 RESULTS

For each of the three methods the sum total of population for Hamlets A, X, and Z was compared against the 'gold standard'.

3.1. Interpolation

The results of the surveys are presented for three different conditions: counting all households, including those crossing the circle circumference (Table 6.a), counting all households inside the circumference and only 50% of the population of the households crossing the circumference (Table 6.b), and counting only the households inside the circumference (Table 6.c).

Table 6.a Unadjusted interpolation from survey (100% of on border cases considered)

Hamlet/ Polygon	Random samples	Kriging std. RMS	Estimate 30m	Estimate 35m	Estimate 40m	Estimate 45m	Estimate 50m
A	set 1	1.333	1148	1141	1202	1288	1226
A	set 2	0.9049	1555	1551	1622	1714	1662
X	set 1	0.6475	781	833	841	846	813
X	set 2	0.9773	740	760	783	788	779
Z	set 1	1.089	1058	1100	1044	1025	1082
Z	set 2	1.009	1138	1181	1115	1100	1166

Hamlet/ Polygon	Random samples	True	Interpolation %
A	set 1	878	130% - 147%
A	set 2	878	177% - 195%
X	set 1	524	149% - 161%
X	set 2	524	141% - 150%
Z	set 1	781	131% - 141%
Z	set 2	781	141% - 151%

Table 6.b Unadjusted interpolation from survey (50% of on border cases considered)

Hamlet/ Polygon	Random samples	Kriging std. RMS	Estimate 30m	Estimate 35m	Estimate 40m	Estimate 45m	Estimate 50m
A	set 1	1.209	945	933	981	1064	1013
A	set 2	0.9209	1122	1127	1177	1237	1201
X	set 1	0.6796	469	501	506	506	482
X	set 2	0.999	586	600	619	623	616
Z	set 1	1.851	857	891	840	827	884
Z	set 2	1.04	873	906	852	844	898

Hamlet/ Polygon	Random samples	True	Interpolation %
A	set 1	878	106% - 121%
A	set 2	878	128% - 141%
X	set 1	524	90% - 97%
X	set 2	524	112% - 119%
Z	set 1	781	106% - 114%
Z	set 2	781	108% - 116%

Table 6.c Unadjusted interpolation from survey (0% of on border cases considered)

Hamlet/ Polygon	Random samples	Kriging std. RMS	Estimate 30m	Estimate 35m	Estimate 40m	Estimate 45m	Estimate 50m
A	set 1	1.094	726	710	752	820	784
A	set 2	0.9457	713	724	742	774	759
X	set 1	0.9678	179	193	191	191	184
X	set 2	1.176	401	421	422	431	423
Z	set 1	1.06	700	733	676	660	725
Z	set 2	1.028	607	631	589	589	630

Hamlet/ Polygon	Random samples	True	Interpolation %
A	set 1	878	81% - 93%
A	set 2	878	81% - 88%
X	set 1	524	34% - 37%
X	set 2	524	77% - 82%
Z	set 1	781	84% - 94%
Z	set 2	781	75% - 81%

It can be seen from Table 6.a that the inclusion of all border cases gives rise to over-estimation of the true values on all counts of interpolation for all the grid spacings (30m, 35m, 40m, 45m, 50m). Table 6.c shows the other extreme that excluded all border cases, giving rise to under-estimation of the true values across the board. It appears that a 50 percent inclusion of border cases yielded the best possible estimated values between 90 and 121 percent of the true values. It also appears that 30m-grid spacing tended to produce the best estimates in most cases.

3.2 T-square

The T-square method was carried out in 7 hamlets. The total number of random points was proportionally distributed in each hamlet according to the surface area of each hamlet. The table below shows the results for each hamlet and the sum of all of them in respect the total population.

Table 7. T Square method vs. 'gold standard'

	<5					>=5				
Area	LL	Estimate	UL	TRUE	TSQUARE %	LL	Estimate	UL	TRUE	TSQUARE %
A	58	74	103	134	55.50	258	329	456	744	44.27
B	18	28	64	63	43.95	40	62	143	142	43.87
D	11	18	40	67	26.17	45	70	162	222	31.59
E	158	248	570	71	348.78	290	454	1046	187	242.78
F	71	111	256	54	205.47	218	341	786	182	187.58
X	39	52	78	55	94.41	301	402	605	469	85.62
Z	76	103	160	98	104.91	579	786	1224	683	115.11
		633		**542**	**116.77**		**2445**		**2629**	**93.00**

Total				
LL	Estimate	UL	TRUE	TSQUARE %
316	404	559	878	45.98
57	90	207	205	43.90
56	88	202	289	30.34
448	702	1616	258	271.95
289	452	1042	236	191.67
339	453	683	524	86.54
655	889	1384	781	113.83
	3078		**3171**	**97.06**

3.3. Quadrat

The Quadrat method was carried out in the same manner as the T-square. The tables below show the results of the population estimation using the Poisson distribution and the Negative Binomial distribution.

Table 8.a Quadrat method vs. 'gold standard' (Poisson)

	<5					>=5				
Area	LL	Estimate	UL	TRUE	QUADRAT %	LL	Estimate	UL	TRUE	QUADRAT %
A	0	51	163	134	38.06	251	1276	2298	744	171.51
B	-	0	-	63	-	-	-	-	142	-
D	0	185	980	67	276.12	0	74	392	222	33.33
E	0	146	403	71	205.63	0	438	884	187	234.22
F	0	83	208	54	153.70	94	291	489	182	159.89
X	0	0	0	55	0.00	0	165	581	469	35.18
Z	0	0	0	98	0.00	249	867	1479	683	126.94
		465		**542**	**85.79**		**3111**		**2629**	**118.33**
ALL	0	**722**	**1625**		**133.14**	**2495**	**3735**	**4976**		**142.09**

Total				
LL	Estimate	UL	TRUE	QUADRAT %
307	1326	2344	878	151.03
-	-	-	205	-
0	259	1372	289	89.62
69	584	1099	258	226.36
164	375	585	236	158.90
0	165	581	524	31.49
249	864	1479	781	110.63
	3573		**3171**	**112.68**
3141	**4457**	**5773**		**140.56**

Table 8.b Quadrat method vs. 'gold standard' (Negative Binomial)

	<5					>=5				
Area	LL	Estimate	UL	TRUE	QUADRAT %	LL	Estimate	UL	TRUE	QUADRAT %
A	~	51	~	134	38.05	40	1116	10477	744	150.01
B	-	-	-	63	-	-	-	-	142	-
D	0	110	40502225	67	164.45	0	54	18488	222	24.48
E	2	139	369	71	196.39	0	291	16009	187	155.87
F	65	108	560	54	200.08	3	227	1100	182	124.49
X	0	0	0	55	0.00	9	140	4491	469	29.83
Z	0	0	0	98	0.00	23	727	3698	683	106.43
		409		**542**	**75.40**		**2555**		**2629**	**97.20**
ALL	**147**	**639**	**13833**		**117.87**	**538**	**3259**	**7110**		**123.95**

Total				
LL	Estimate	UL	TRUE	QUADRAT %
38	1150	11388	878	130.96
-	-	-	205	-
0	146	344818392	289	50.62
0	371	22098	258	143.97
2	262	1506	236	110.89
9	140	4491	524	26.70
23	727	3698	781	93.08
	2796		**3171**	**88.18**
488	**3615**	**8473**		**114.01**

3.4. Comparing three methods (Interpolation, Quadrat and T-Square)

It can be seen from Table 9 below that the interpolation method seems to produce highly accurate values for the 30m-grid spacing (at 104 percent of the true value). In the worst-case scenario, interpolation produced estimates at 124 percent of the true value. These results are quite comparable to those of the Quadrat (at 92 to 108 percent) and T-Square (at 80 percent).

Table 9. Comparing results of the three methods

Total population of Hamlets A + X + Z =			2183		
Interpolation	A	X	Z	A+X+Z	Accuracy
Set1 30m	945	469	857	2272	104%
Set2 30m	1122	586	873	2580	118%
Quadrat	A	X	Z	A+X+Z	Accuracy
Poisson	1326	165	864	2355	108%
Negative Binomial	1150	140	727	2017	92%
T-Square	404	453	889	1746	80%

3.5. Computer Simulation

A comparison of Tables 6.b and 10 shows that the results were comparable, although population estimates from the field survey showed lesser variation.

Table 10. Computer simulated results using the same sets of random points for field survey

Hamlet/ Polygon	Random samples	Kriging std. RMS	Estimate 30m	Estimate 35m	Estimate 40m	Estimate 45m	Estimate 50m
A	set 1	1.115	1076	1072	1125	1198	1151
A	set 2	1.372	650	644	682	705	701
X	set 1	0.9656	324	337	340	333	332
X	set 2	1.278	383	411	405	427	417
Z	set 1	1.947	748	782	745	731	767
Z	set 2	1.026	579	601	564	562	601

Hamlet/ Polygon	Random samples	True	Interpolation %
A	set 1	878	122% - 136%
A	set 2	878	73% - 80%
X	set 1	524	62% - 65%
X	set 2	524	73% - 81%
Z	set 1	781	94% - 100%
Z	set 2	781	72% - 77%

A further 20 sets of random points were generated for Hamlet A and 10 sets each for hamlets X and Z using the random point generator in ArcView 3.1. For hamlet A the average values of the 20x5 simulations are between 94 to 106 percent of the true value with the population estimate at 40 m grid spacing yielding the best result. With reference to hamlet X the average values of the 10x5 simulations are between 106 and 112 percent of the true value with the population estimate at 30m-grid spacing yielding the best result. Finally for hamlet Z the average values of the 10x5 simulations are between 90 to 96 percent of the true value with the population estimate at 35m-grid spacing yielding the best result.

Further examination of the estimated values for the three hamlets (averages of estimated values from different grid spacing for each set of random points) against the RMS Errors was also done, and the results are shown in Figure 2. Hamlets A and Z show a clustering of the estimated values about the true value and within acceptable ranges of RMS errors, whilst for hamlet X the RMS Errors of the simulations are more widespread. The results are nonetheless inconclusive and further tests/simulations are necessary.

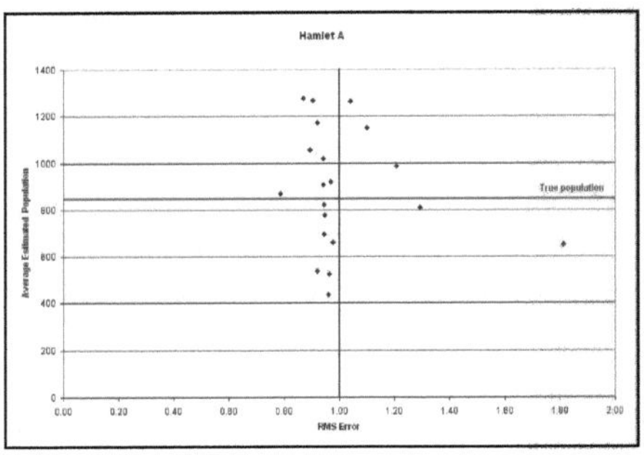

(a) Hamlet A

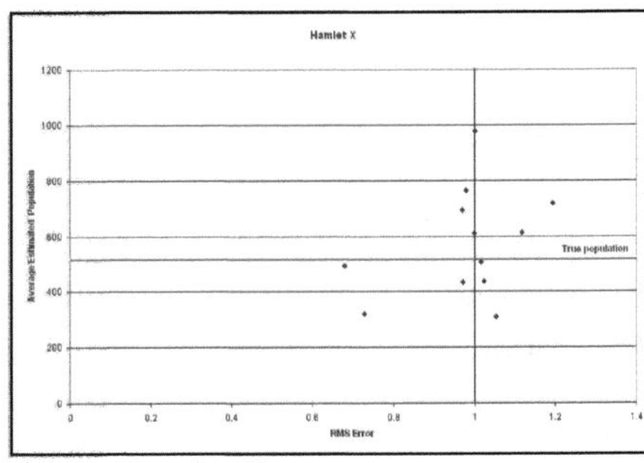

(b) Hamlet X

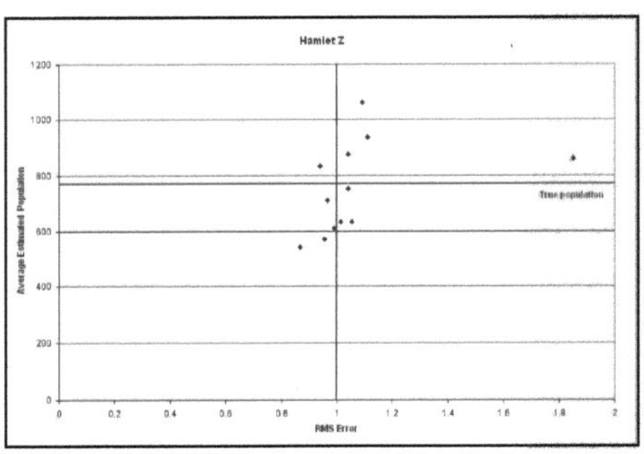

(c) Hamlet Z

Fig. 2. Graphic plots of Average Estimated Population against RMS Error

4 Conclusions And Recommendations

Our findings are encouraging although not conclusive. In our assessment of the feasibility of the interpolation methods, we noted the following:

- Spatial interpolation was as easy to implement in the field as traditional sampling methods in this setting.
- The guideline of 1 random point per 150x150m^2 area (subject to a minimum of 10 sampled points regardless of total area) yielded a sufficient number of sampling locations for the interpolation.

- The guideline of 25m-search radius about the random sampling points produced acceptable results.
- Through computer simulations, the interpolation method seems to generate consistent results within a closed area but the precision of the population estimate is subject to the location of the sampling points (although the randomness of the distributional pattern of sampling locations does not seem to have a bearing on accuracy).
- The grid spacing (30m, 35m, 40m, 45m, 50m) in surface interpolation does not seem to have a significant impact on the precision of the population estimates.
- Our findings are not conclusive because the homogeneity of settlement distribution is violated.
- In comparing the performances among the three estimation methods (spatial interpolation, Quadrat, and T-Square), we made the following observations:
- The field test showed all three methods feasible in practice and the interpolation method to be easiest to undertake except when border cases need to be determined. It is uncertain, however, whether the method would apply with the same relative ease in areas of dense settlement.
- The results of the methods were tested statistically and verifiable.
- It was not possible with this study to declare operational efficiencies (savings in time or human resources) of the interpolation method against the Quadrat and T-Square methods because of standardization problems in the data collection process. However the view of the field teams was that the interpolation method was the most efficient method of data collection: the circle was easier to construct than a quadrat, and less were needed. The interpolation method was also far less intrusive on the members of the households than the T-square.
- In terms of use of the use of technology applied to the three methods, the training of field workers in GPS and survey method is probably the same for all 3 methods, and the local field workers picked this up very quickly. However there is a need for some computer literacy to generate sample points and download into the GPS, again not that different for all 3 methods.
- In terms of generating the results, at the moment the Quadrat and T-square are quick and robust due to the Excel application developed to simplify the data analysis. The interpolation still requires a specialized competence in GIS until a user-friendly application has been developed.

4.1 Recommendations and Future Research

It appears that the interpolation method has great potential given its ease of operation in the field. More tests on the method are desirable to address specific concerns listed below.

- The interpolation method needs to be repeated in more homogeneous settings.
- The sampling radius of 25m needs to be tested in areas of dense distribution.
- The consideration of border cases must be more clearly defined.
- The delineation of the study area by the initial mapping using GPS reception is to be standardized (e.g. 5-10m from the edge of shelters bordering the settlement).
- The relationship between sampling radius and sensitivity of grid spacing in interpolation needs further exploration.
- The use of other kriging methods (e.g. universal kriging) in surface interpolation should be investigated.
- A new field test should be carried out using again a gold standard and counting the household population during the field test data collection
- A user-friendly application for field data analysis may be envisaged after further field tests.

Acknowledgements

This research has been the result of a close and fruitful partnership among different institutions, namely: World Health Organization, Epicentre, University of Hong Kong, Department of Geography; University of Sheffield, Department of Geography, UK; University of Mahidol, Bangkok, that have contributed to the success of the field activities and the production of the final results.

A special thanks and acknowledgement should go to Prof Pratap Singhasivanon, Dean of the Faculty of Tropical Medicine at Mahidol University in Bangkok, and all his staff who supported all field activities in Suanphung. He hosted the whole team in the field at the Rajanagarindra Tropical Disease International Centre providing all facilities including transport for the field research.

References

[1] Brown V, Coulombier D, Belanger F, Jacquier G, Balandine S, Legros D (2001) Rapid assessment of population size by area sampling in disaster situations. *Disasters* Vol 25(2): 164-17

[2] Croisier A (2003) *Draft research protocol estimating population's size in emergency situation.* WHO/CSR/Lyon January 2003 (unpublished)

[3] Cromley KE, Mclafferty LS (2002) GIS and Public Health. The Guilford Press, pp 198-204

[4] Espie E (2000) *Rassemblement exceptionnels de populations: Evaluation de trois methods d'estimation de populations.* Memoire de DEA « Epidemiologie et Interventions en Santé publique ». Université V. Segalen – Bordeaux II, (unpublished).

[5] ESRI (2001) Introduction to Surface Analysis. http/siteus.esri.com/courses/spatial analysis

[6] Grais RF, Coulombier D, Ampuero J, Lucas ES, Barretto AT, Jacquier G, Diaz F, Balandine S, Mahoudeau C, Brown V (2006) Are rapid population estimates accurate? A field validation of two different rapid population assessment methods. *Disasters* (in press)

[7] Ludwig JA, Reynolds JF (1988) Statistical Ecology: A Primer on Methods and Computing, Wiley-Interscience Publication

[8] Noji EK (2005) Estimating population size in emergencies, Editorial, *Bulletin of The World Health Organization*, 83 (3)

[9] Tellford J, Gibbons L, Van Brabant K (1997) Counting and identification of beneficiary populations in emergency operations: registration and its alternatives. London: Overseas Development Institute

Avian Influenza Outbreaks of Poultry in High Risk Areas of Thailand, June-December 2005

K. Chanachai[a], T. Parakgamawongsa[b], W. Kongkaew[a], S. Chotiprasartin-thara[c] and C. Jiraphongsa[a]

[a] Field Epidemiology Training Program, Bureau of Epidemiology, Thailand
[b] Suphanburi Provincial Livestock Office, Department of Livestock Development, Thailand
[c] Bureau of Animal Disease Control and Veterinary Services, Department of Livestock development, Thailand

Abstract: There were 848 suspected and 188 confirmed Avian Influenza (AI) in poultry flocks in Thailand in 2005. Between June to December 2005, the Suphanburi province reported the highest number of AI confirmed flocks. This chapter is an account of the outbreak with a description of risk factors and recommended preventive measures in poultry farming. Our case definition of a suspected flock is a poultry flock with abnormal deaths of more than 10% within one day or more than 40% within three days, while an AI confirmed flock is a suspected flock with laboratory confirmation. We interviewed poultry owners and conducted a case control study in the Suphanburi province in which 25 of 79 reported cases of suspected flocks were confirmed between June to December 2005. Most of the confirmed flocks (about 64%) were backyard native chickens. The percentage of deaths in the first three days for backyard native chicken and free-grazing duck flocks were 19% compared to 0.46% for layer and broiler flocks. A number of risk factors, such as contact with waterfowl or proximity to feeding areas, was associated with the AI infection. The association between the number of poultry population and that of AI confirmed flocks by areas was also explored. Other than improving bio-security in poultry farms, some practical recommendations including how to deal with dead birds and how to protect poultry from neighboring infected poultry should be communicated especially to backyard chicken owners.

Keywords: Avian influenza, H5N1, poultry, Thailand

1 Background

Avian influenza is in the group of Influenza A virus, family Orthomyxoviridae. Various serotype of Avian Influenza virus was found dependent on Neuraminidase and Hemagglutinin. H5N1 Avian influenza (AI) is a threat to the global poultry industry and a cause of increased global awareness of pandemic influenza in human. AI had its first human death reported in Hong Kong in 1997 (MMWR, 1997). So far, it has spread widely into many regions and resulted in human infections in many countries including Azerbaijan, Cambodia, China, Djibouti, Egypt, Indonesia, Iraq, Thailand, Turkey, and Viet Nam (WHO 1997).

Thailand reported the first AI outbreak in poultry in January 2004 (Tiensin et al. 2005). Thereafter, epidemics in poultry in Thailand occurred periodically. Between 2004 and 2005, thousands of reported outbreaks in poultry flocks were confirmed with AI and most of the infected flocks were located in specific areas in the middle and lower northern parts of Thailand (Tiensin et al. 2005). AI outbreaks occurred continuously from the end of 2004 and the first period of epidemic ended in April 2005. Three month later in July 2005, an outbreak occurred again in the Suphanburi province located in the middle part of Thailand and more sporadic outbreaks were reported in various provinces soon after. 70 flocks were found positive for AI between July and December 2005. Provinces with higher numbers of confirmed flocks included Suphanburi (25 flocks), Kampangphet (23 flocks), and Kanchaburi (6 flocks) (Figure 1).

Risk factors of AI occurrence between July 2004 and May 2005 were studied. Gilbert et al. (2006) concluded that wetland areas and free-grazing ducks were associated with AI occurrences. No further study concerning the risk factors of AI was done after May 2005. Although outbreak investigations were conducted by local officers at each occurrence, risk factors of AI among the poultry flocks have remained unclear. In an attempt to better understand the epidemiology of AI epidemic, we conducted an investigation on infected flocks in the Suphanburi province to characterize AI outbreaks, identify potential risk factors for the outbreak, and construct recommendations for the prevention and control of similar outbreaks in the future.

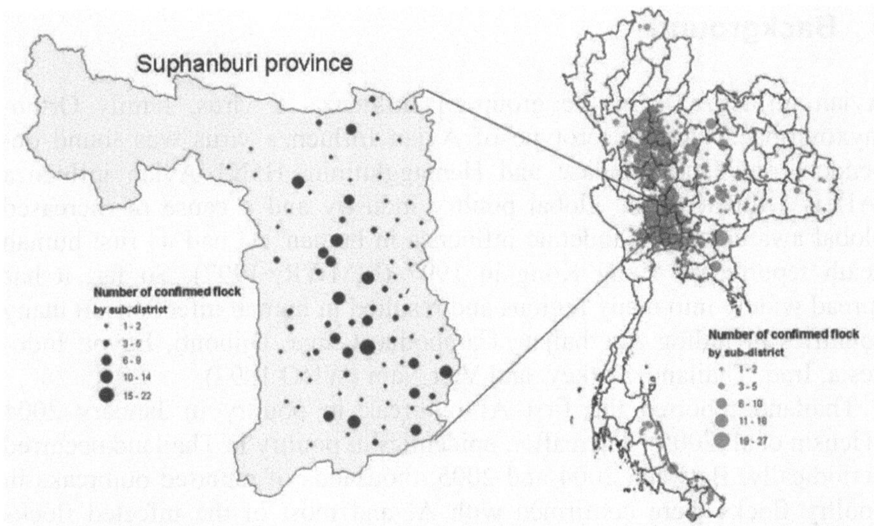

Fig. 1. Geographical location of AI outbreaks in poultry flocks between July 2004 and December 2005, Thailand

2 Methodology

Our study area was limited to the Suphanburi province because it was the most effected area during the study period of June-December 2005. Our case study may not be representative of AI outbreaks for the whole country. However, the study is intended to draw some important epidemiological characteristics and possible risk factors of AI. We reviewed positive cases reported in Suphanburi from the Department of Livestock Development in Thailand, as well as results of active case findings during the outbreak in poultry from the Suphanburi Provincial Livestock Office. Poultry owners of all confirmed flocks in the province were interviewed about clinical symptoms, onset date and numbers of dead poultry in the first three days and other characteristics of the outbreak.

We modified case definitions in this investigation from that used by the Department of Livestock Development at the time. In this study, suspected flocks included any poultry flocks with more than 10% deaths within one day from unknown causes and had at least one of following clinical signs: sudden death, respiratory system illness, neurological system illness or diarrhea. Confirmed flocks would include poultry flocks with laboratory confirmation of H5 by hemagglutination test performed by the National Institute of Animal Health in Thailand.

A retrospective matched case-control study among confirmed AI flocks was done in November 2005 to identify potential risk factors of AI outbreaks in poultry flocks. The rate of case to control was one to two. The cases were confirmed flocks in our descriptive study. Matched Controls were randomly selected from within 20 households which had no abnormal deaths of poultry between June and November 2005 around a case household. Hypotheses of possible risk factors were generated from literature reviews and preliminary investigations by local officers. We defined several variables for proving possible risk factors, as shown in Table 1. Matched odds ratios were calculated for implicated risk factors, and tested for significance at 0.05 using the Chi-square statistic. A multivariate matched analysis was conducted by conditional logistic regression using Epi Info version 3.3.2 (CDC, Atlanta, Georgia) to determine adjusted odds ratio (OR).

Data on poultry population surveyed in July 2005 by the provincial livestock office and results of AI survey in natural birds by the provincial natural resources authority were also reviewed. The association between poultry population and the numbers of AI suspected and confirmed flocks at the village level was analyzed using correlation analysis.

Table 1. Hypotheses of risk factors of AI in poultry and related variables used in outbreak investigation in Suphanburi province between June and December 2005

Risk factors	Variables
Virus was introduced from neighboring infected poultry flocks to confirmed flocks through shared feeding areas.	Having abnormal deaths of neighboring poultry from activities with other domestic poultry in shared feeding areas
Virus was introduced from dead or sick poultry moved into farm sites by various routes	Introduction of dead or sick poultry into farm sites
Free-grazing ducks introduced virus to poultry flocks through contact or in close proximity with infected flocks	Contact or proximity with free-grazing ducks
Water fowls introduced virus to poultry flocks through contact or in close proximity with infected flocks	Contact or proximity with water fowl feeding areas
Mechanical transmission of virus via fighting cocks or via owners who had activities in cock fighting courtyards	Cock fighting activities
Mechanical transmission of virus via movement of healthy poultry into farm sites or feed containers	Introduction of possible fomites

3 Results

Between June and December 2005, 54 suspected and 25 confirmed flocks were reported by the District Livestock Officer in Suphanburi province. These confirmed flocks were found in 22 sub-districts and 24 villages. The spatial and temporal distribution of suspected and confirmed flocks are shown in Figure 2.

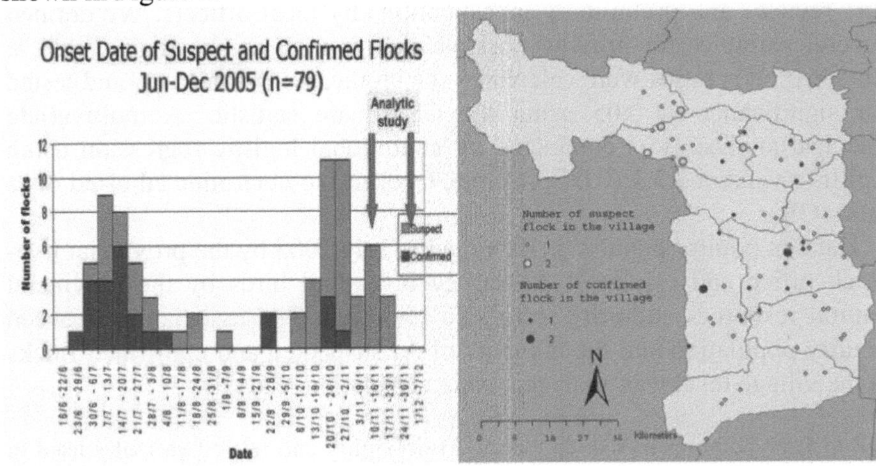

Fig. 2. Spatial and temporal distribution of suspected and confirmed AI flock in Suphanburi province, June to December 2005

The confirmed flocks included 16 backyard chicken flocks, all of which was for non-commercial purposes. There were also three quail flocks, two free-grazing duck flocks, two layer flocks and two broiler flocks which were all for commercial purposes. The largest number of livestock per flock was the quail flocks primarily raised for egg production (Table 2).

Table 2. Flock Size by types of AI confirmed flocks in Suphanburi province, June- December 2005

Type of flock	Number (%)	Median Flock Size (range)
Backyard Chickens	16 (64)	100 (10-200)
Free-grazing ducks	2 (8)	3,450 (2,400-4,500)
Layers	2 (8)	1,010 (19-2,000)
Broilers	2 (8)	9,400 (8,000-10,800)
Quail	3 (12)	96,000 (50,000-122,000)
Total	25 (100)	-

The cumulative mortality rates in the first three days of backyard and free-grazing duck flocks were 4, 10 and 19 percents respectively, while the cumulative mortality rates in the first three days of broiler, layer and quail were 0.1, 0.3 and 0.5 percents respectively. Rates for the latter were very low compared with those among backyard and free-grazing duck flocks. The timeliness of important events among the confirmed flocks was measured in terms of the median time needed to report abnormal deaths of poultry from the owner to the responsible persons-in-charge: (1) median time to the officer or the headman of a village was two days, with a range of zero to nine days; and (2) median time from owners to inform the responsible person to cull poultry flocks was two days, with a range of zero to 18 days.

From the univariate matched analysis, we found association between AI and (i) abnormal deaths of neighboring poultry and (ii) contact or proximity with waterfowl feeding areas. However, no association was found between AI and (i) contact or proximity with free-grazing ducks, (ii) cock fighting activities, and (iii) introduction of possible fomites (Table 3). The multivariate matched analysis by multiple logistic regressions found association between AI and contact with waterfowl or proximity with their feeding areas (Table 4).

Table 3. Results of possible risk factors by univariate matched analysis

Risk factors	Case		Control		Matched OR (95% CI)	P-value
	Exp	Non-exp	Exp	Non-exp		
Abnormal deaths of neighboring poultry through shared feeding areas with other domestic poultry	6	19	1	49	12.00 (1.44-99.68)	0.01*
Introduction of dead poultry to farm sites (by dogs)	3	22	0	50	N/A	N/A
Contact or proximity with free-grazing ducks	2	23	6	44	0.67 (0.13-3.30)	0.38
Contact or proximity with water fowl feeding areas	15	10	14	36	6.33 (1.64-24.47)	0.008*
Cock fighting activities	4	21	12	38	0.60 (0.17-2.10)	0.26
Introduction of possible fomites (vehicles, poultry)	6	19	8	42	2.00 (0.48-8.35)	0.58

* significant at 0.01 and 0.05

Table 4. Results of possible risk factors with ∞ value > 0.20 by multivariate matched analysis

Risk factors	Case		Control		Matched OR (95%CI)	P-value
	Exp	Non-exp	Exp	Non-exp		
Contact or proximity with water fowl feeding areas	15	10	14	36	4.01 (1.04-15.45)	0.04*
Abnormal deaths of neighboring poultry through shared feeding areas with other domestic poultry	6	19	1	49	7.15 (0.80-63.83)	0.08

* significant at 0.05

A survey of natural birds in the province was conducted between July and December 2006. 399 natural live birds were captured for AI testing at the Veterinary Faculty of Mahidol University in Bangkok. The results of all samples were negative. However, another AI survey of natural birds by the natural resource authority during the same period found AI positive in birds (including sparrow, pigeon and the Asian Open-bill Stork) in two neighboring provinces of Suphanburi province (Aung Thong and Kanchanaburi provinces) that also had AI outbreaks in poultry. Nonetheless, correlation analyses between numbers of AI suspected and confirmed flocks in villages and numbers of total free-grazing duck flocks, backyard poultry flocks, layer flocks or broiler flock in Suphanburi by village showed no association with the disease.

4 Discussions

We found a lower mortality rate of poultry flocks in commercial farms compared with backyard and free-grazing duck flocks. This result may not represent the pathogenicity of AI virus itself because it may be due to others factors; for example, factors related to owner behaviors such as greater alertness of the commercial poultry owners to unusual deaths or illnesses of their poultry which may lead to early detection of the disease. However, our finding of a low mortality rate in infected flocks is beneficial to the construction of case definitions in AI surveillance system. A low mortality rate in case definition is needed for early disease detection especially in commercial farms. Our result which contained lower mortality supports the new case definition recommended by the FAO, (FAO, 2004). Although the median period of two days (i.e. from the occurrence of dead poultry to reporting to responsible persons and from owner reporting to culling poultry) seemed satisfying, some cases had prolonged reporting and culling times which may lead to epidemics of a wider coverage. Increased disease awareness and effectiveness of surveillance networks are key measures to tackling this problem. Thus, it is recommended that poultry owners are made aware of the disease and able to recognize AI such that responsible persons could be informed of possible infection as early as possible, e.g. within 24 hours upon detection of abnormalities in their poultry. Surveillance networks should be also strengthened to facilitate early detection of AI in poultry. Moreover, an active AI control by culling suspected flocks without waiting for laboratory results may be enforced to minimize disease spreading when abnormality is notified by poultry owners.

There was epidemiological evidence of the waterfowl as a risk factor of AI. Our empirical findings showed that flocks in farms close to waterfowl feeding areas or had presence of waterfowl in poultry flock areas had a greater chance of contracting the AI virus from the waterfowl. Even so, wild bird samples from the natural bird survey program in Suphanburi showed negative results for AI, although some wild birds including waterfowl in the neighboring provinces were found positive for AI but without any clinical signs. We thus conclude that waterfowl has the potential to enhance spreading of the disease as mentioned in other publications (Normile 2006, Sims et al. 2005, Werner 2006) which is also supported by empirical findings of this study.

It is unclear as to whether virus transmission from the waterfowl to domestic poultry occurred by mechanical or biological means. The presence of free-grazing ducks surrounding the confirmed flocks was found not associated with AI in this study, even though free-grazing ducks can be a natural reservoir for AI virus (Tumpey et al. 2003, Lee et al. 2005, Chen et al. 2004) and it was proven a risk factor in Thailand in another period (Gilbert et al. 2006). This observation may suggest that the characteristics of AI outbreaks in our study period may be different from outbreaks in the past. It might be because the Department of Livestock Development implemented movement controlling measures on free-grazing ducks during the study period which we found rather effective through our field investigation.

Abnormal death of neighboring poultry in flocks sharing feeding areas with other domestic poultry was a factor of borderline significance. This finding is important to draw the recommendation for poultry owners to keep their poultry isolated from other poultry flocks. Movement of dead poultry or carcasses into the three farm sites was found an act of dogs. Although we did not include this risk factor in the multivariate analysis because none of the flocks in the control group had a history of moving in sick or dead poultry, it is still reasonable to recommend owners to avoid moving in sick poultry into their farms through intentional or unintentional acts. Fences for example can be erected to separate other animals (e.g. dogs and cats) from poultry farms or disinfection areas with sick or dead poultry within the farm site.

We considered recall bias a limitation of our study which could occur because of different degrees of awareness between the owners of case and control groups. Misclassification of case and control could occur because we did not conduct serological tests among the control flocks. However, a number of research indicated that sub-clinical infection in chicken was rare (Swayne and Pantin-Jackwood 2006, Werner 2006) and thus this limitation should be minimal. Further studies are needed for a better understanding in the epidemiology of AI; for example, the situation of AI in wild birds and its association with AI outbreaks, and factors affecting case reporting of AI occurrences by poultry owners. In conclusion, we found a low mortality rate of poultry flocks in the first three days after onset. Contact with waterfowl or proximity to waterfowl feeding areas was found a risk factor of AI occurrences in Suphanburi province between June and December 2005.

In terms of prevention of AI in the future, it is critical that other practical preventive measures in backyard flocks be strengthened besides strengthening of bio-security in commercial poultry farms. These measures include keeping poultry away from the waterfowl and also from neighboring poultry, increasing AI awareness among owners and strengthening surveillance networks in the local communities.

Acknowledgement

We would like to acknowledge staff and trainees of the Field Epidemiology Training Program-Thailand, The Bureau of Animal Disease Control and Veterinary Services, Department of Livestock Development, Suphanburi Provincial Livestock Office for their contribution in this investigation. Special thanks are extended to ESRI-Thailand, Dr. Chris Skelly, Dr. Micheal O'Reilly and Dr. Poh-Chin Lai for their support in the preparation of the manuscript.

Reference

[1] Chen H, Deng G, Li Z, Tian G, Li Y, Jiao P, Zhang L, Liu Z, Webster RG, Yu K (2004) The evolution of H5N1 influenza viruses in ducks in southern China. Proc Natl Acad Sci U S A. 101: 10452–10457

[2] FAO (2004). Guiding principles for highly pathogenic avian influenza surveillance and diagnostic networks in Asia. http://www.fao.org/docs/eims/upload//210749

[3] Gilbert M, Chaitaweesub P, Parakamawongsa T, Premashthira S, Tiensin T, Kalpravidh W, Wagner H, Slingenbergh J (2006) Free-grazing ducks and highly pathogenic avian influenza, Thailand. Emerg Infect Dis 12:227-34

[4] Lee CW, Suarez DL, Tumpey TM, Sung HW, Kwon YK, Lee YJ, Choi JG, Joh SJ, Kim MC, Lee EK, Park JM, Lu X, Katz JM, Spackman E, Swayne DE, Kim JH (2005) Characterization of highly pathogenic H5N1 avian influenza A viruses isolated from South Korea. J Virol 79:3692-702

[5] Normile D (2006) Avian influenza. Evidence points to migratory birds in H5N1 spread. Science 311: 1225

[6] Sims LD, Domenech J, Benigno C, Kahn S, Kamata A, Lubroth J, Martin V, Roeder P (2005) Origin and evolution of highly pathogenic H5N1 avian influenza in Asia. Vet Rec 157: 159-64

[7] Swayne, DE and Pantin-Jackwood M (2006). Pathogenicity of avian influenza viruses in poultry. Dev Biol (Basel) 124: 61-7

[8] Tiensin T, Chaitaweesub P, Songserm T, Chaisingh A, Hoonsuwan W, Buranathai C, Parakamawongsa T, Premashthira S, Amonsin A, Gilbert M, Nielen M, Stegeman A (2005) Highly pathogenic avian influenza H5N1, Thailand, 2004. Emerg Infect Dis 11: 1664-72

[9] Tumpey TM, Suarez DL, Perkins LE, Senne DA, Lee J, Lee YJ, Mo IP, Sung HW, Swayne DE (2003) Evaluation of a high-pathogenicity H5N1 avian influenza A virus isolated from duck meat. Avian Dis 47(3 Suppl): 951-5

[10] Werner O (2006) Highly pathogenic avian influenza--a review. Berl Munch Tierarztl Wochenschr 119: 140-50

[11] WHO (1997) Isolation of avian influenza A(H5N1) viruses from humans--Hong Kong, May-December 1997. MMWR Morb Mortal Wkly Rep 46: 1204-7

Contact Information

Exploratory Spatial Analysis Methods in Cancer Prevention and Control **2**

Gerard Rushton
The University of Iowa
Department of Geography
316 Jessup Hall
The University of Iowa
Iowa City, IA 52242
Email: gerard-rushton@uiowa.edu

Environmental Risk Factor Diagnosis for Epidemics **15**

Jin-feng Wang
State Key Laboratory of Resources and Environmental Information System
Institute of Geographical Sciences and Natural Resources Research
Chinese Academy of Sciences, China
A11, Datun Road, Beijing 100101, PR China
Email: wangjf@Lreis.ac.cn

A Study on Spatial Decision Support Systems for Epidemic Disease Prevention Based on ArcGIS **30**

Kun Yang
Faculty of Tourism and Geographic Science, Yunnan Normal University, No.298, Dec. 1st Street, Kunming, 650092, China
Email: kmdcynu@163.com

Yan-bo Cao
Faculty of Tourism and Geographic Science, Yunnan Normal University, Kunming, China.

Quan-li Xu
Faculty of Tourism and Geographic Science, Yunnan Normal University, Kunming, China.

Shung-yun Peng
Faculty of Tourism and Geographic Science, Yunnan Normal University, Kunming, China.

Development of a Cross-Domain Web-based GIS Platform to Support Surveillance and Control of Communicable Diseases **44**

Cheong-wai Tsoi
Lands Department, Hong Kong SAR, China
23/F North Point Government Offices,
333 Java Road, North Point, Hong Kong
Email: cwtsoi@gmail.com

Abdulkader Murad
King Abdulaziz University, Jeddah, Saudi Arabia
Email: amurad25@hotmail.com

Ester Cerin
Institute of Human Performance,
The University of Hong Kong, Hong Kong SAR, China
Email: ecerin@hku.hk

Eva Leslie
School of Health and Social Development,
Deakin University, Australia

Adrian Baumand
NSW Centre for Physical Activity and Health,
University of Sydney, Australia

Neville Owen
Cancer Prevention Research Centre,
University of Queensland, Australia

Eva Leslie
School of Health and Social Development,
Deakin University,
Waterfront Campus,
1 Gheringhap St,
Geelong, Australia
Email: evie.leslie@deakin.edu.au

Ester Cerin
Institute of Human Performance,
University of Hong Kong, HK,
China

Lorinne duToit
Cancer Prevention Research Centre,
University of Queensland, Australia

Neville Owen
Cancer Prevention Research Centre,
University of Queensland, Australia

Adrian Bauman
NSW Centre for Physical Activity and Health,
University of Sydney, Australia

Developing Habitat-suitability Maps of Invasive Ragweed (Ambrosia artemisiifolia.L) in China Using GIS and Statistical Methods **105**

Hao Chen[a,b]
[a] School of Remote Sensing Information Engineering,
Wuhan University, Wuhan, China
[b] National Geomatics Center of China, Beijing, China
Email: chenhao087@sohu.com

Lijun Chen
National Geomatics Center of China,
Beijing, China,

Thomas P. Albright
Department of Zoology,
University of Wisconsin-Madison,
WI, USA

An Evaluation of a GIS-aided Garbage Collection Service for the Eastern District of Tainan City **122**

Jung-hong Hong
Department of Geomatics
National Cheng-Kung University
Tainan, Tai-wan
Email: junghong@mail.ncku.edu.tw

Yue-cyuan Deng
Department of Geomatics
National Cheng-Kung University
Tainan, Tai-wan

A Study of Air Quality Impacts on Upper Respiratory Tract Diseases **142**

Huey-hong Hsieh
Department of Environment and Resources Engineering
Diwan College of Management, Taiwan
Email: hhsieh@dwu.edu.tw

Bing-fang Hwang
School and Graduate Institute of
Occupational Safety and Health
College of Public Health,
China Medical University, Taiwan

Yu-ming Wang
Department of Information
Management,
Diwan College of Management,
Taiwan

Shin-jen Cheng
Department of Environment and
Resources Engineering,
Diwan College of Management,
Taiwan

Franklin F.M. So
Experian Asia Pacific, Experian Limited
Email: fmso@graduate.hku.hk

P.C. Lai
Department of Geography, The University of Hong Kong

Ngai Sze Wong
Geography & Resource Management Department,
Chinese University of Hong Kong,
Hong Kong, China
Email: ngaisiwong@yahoo.com.hk

Chi Yan Law
Geography & Resource
Management Department,
Chinese University of Hong Kong,
Hong Kong, China

Man Kwan Lee
Geography & Resource
Management Department,
Chinese University of Hong Kong,
Hong Kong, China

Shui Shan Lee
Centre for Emerging Infectious
Disease,
Chinese University of Hong Kong,
Hong Kong, China

Hui Lin[a,b]
[a] Geography & Resource
Management Department,
Chinese University of Hong Kong,
Hong Kong, China
[b] Institute of Space and Earth
Information Science,
Chinese University of Hong Kong,
Hong Kong, China

Mandy Y.F. Tang
Lands Department, Hong Kong SAR, China
23/F North Point Government Offices,
333 Java Road, North Point, Hong Kong
Hong Kong SAR, China
Email: gis.mandy@gmail.com

Cheong-wai Tsoi
Lands Department, Hong Kong SAR, China

Socio-Demographic Determinants of Malaria in Highly Infected Rural Areas: Regional Influential Assessment Using GIS **194**

Devi M. Prashanthi
Division of Remote Sensing and GIS
Department of Environmental Sciences
Bharathiar University Post, Coimbatore, Tamil Nadu, 641 046, India
Email: prashanthi_devi@yahoo.co.in

C.R. Ranganathan
Department of Mathematics
Tamil Nadu Agricultural University,
Coimbatore, Tamil Nadu, India

S. Balasubramanian
Division of Remote Sensing and GIS
Department of Environmental Sciences
Bharathiar University Post,
Coimbatore, Tamil Nadu, India

A Study of Dengue Disease Data by GIS Software in Urban Areas of Petaling Jaya Selatan **205**

Mokhtar Azizi Mohd Din
Department of Civil Engineering,
Faculty of Engineering,
University of Malaya,
50603 Kuala Lumpur, Malaysia
Email: mokhtarazizi@um.edu.my

Md. Ghazaly Shaaban
Department of Civil Engineering,
Faculty of Engineering,
University of Malaya,
50603 Kuala Lumpur, Malaysia

Leman Norariza
Department of Civil Engineering,
Faculty of Engineering,
University of Malaya,
50603 Kuala Lumpur, Malaysia

Taib Norlaila
Department of Civil Engineering,
Faculty of Engineering,
University of Malaya,
50603 Kuala Lumpur, Malaysia

A Spatial-Temporal Approach to Differentiate Epidemic Risk Patterns **213**

Tzai-hung Wen
Center for Geographical Information Science,
Research Center for Humanities and Social Sciences,
Academia Sinica, Taipei, Taiwan
Email: liumin@watgislab.ae.ntu.edu.tw

Neal H Lin
Institute of Epidemiology,
College of Public Health,
National Taiwan University,
Taipei, Taiwan

Ming-daw Su
Department of Bioenvironmental
Systems Engineering,
National Taiwan University,
Taipei, Taiwan

Katherine Chun-min Lin
Department of Public Health,
College of Public Health,
National Taiwan University,
Taipei, Taiwan

Chwan-chuen King
Institute of Epidemiology,
College of Public Health,
National Taiwan University,
Taipei, Taiwan

I-chun Fan
Center for Geographical Information
Science,
Research Center for Humanities and
Social Sciences,
Academia Sinica, Taipei, Taiwan

***A "Spatiotemporal Analysis of Heroin Addiction" System for Hong Kong* 228**

Phoebe Tak-ting Pang
Department of Geography & Resource Management,
Chinese University of Hong Kong,
Hong Kong, China
Email: phoebepang@gmail.com

Phoebe Lee
Department of Geography &
Resource Management,
Chinese University of Hong Kong,
Hong Kong, China

Shui-shan Lee
Centre of Emerging Infectious
Disease,
Chinese University of Hong Kong,
Hong Kong, China

Wai-yan Leung
Department of Geography &
Resource Management,
Chinese University of Hong Kong,
Hong Kong, China

Hui Lin[a,b]
a Department of Geography &
Resource Management,
Chinese University of Hong Kong,
Hong Kong, China
b Institute of Space and Earth
Information Science,
Chinese University of Hong Kong,
Hong Kong, China

Ashok Hanjagi
Associate Professor
Department of Geography & Geoinformatics
Bangalore University
Bangalore- 560056 India
Email: drashokhanjagi@yahoo.com

Priya Srihari
Guest Faculty
Department of Geography & Geoinformatics
Bangalore University
Bangalore-56 India

A.S. Rayamane
Professor
Department of Geography & Geoinformatics
Bangalore University
Bangalore-56 India

Paul Harris
NGIS Australia & UN Information Management Service
Banda Aceh, Indonesia
Email: paul.harris@ngis.com.au

Dylan Shaw
UN Information Management Service
Banda Aceh, Indonesia

A. Pinto
World Health Organization,
Lyon, France
Email: Pinto@searo.who.int

V. Brown
Epicentre, Paris, France

K.W. Chan
Department of Geography
The University of Hong Kong
Hong Kong

I.F. Chavez
Mahidol University, Bangkok, Thailand

S. Chupraphawan
Mahidol University, Bangkok, Thailand

R.F. Grais
Epicentre, Paris, France

P.C. Lai
Department of Geography
The University of Hong Kong
Hong Kong

Author Index

Subject Index

This publication is printed with funding support from:

香港特別行政區政府工商及科技局
COMMERCE, INDUSTRY AND TECHNOLOGY BUREAU
THE GOVERNMENT OF THE HONG KONG
SPECIAL ADMINISTRATIVE REGION

MIX
Papier aus verantwortungsvollen Quellen
Paper from responsible sources
FSC® C105338

If you have any concerns about our products,
you can contact us on
ProductSafety@springernature.com

In case Publisher is established outside the EU,
the EU authorized representative is:
Springer Nature Customer Service Center GmbH
Europaplatz 3, 69115 Heidelberg, Germany

Printed by Libri Plureos GmbH
in Hamburg, Germany